Beginner's Guide to Central Heating

Beginner's Guides are available on the following subjects:

Audio
Central Heating
Colour Television
Computers
Domestic Plumbing
Electric Wiring
Electronics
Integrated Circuits
Radio
Tape Recording
Television
Transistors
Woodworking

Beginner's
Guide to
Central
Heating

W. H. Johnson
C.Eng. F.I. Gas E., F.I.S.T.C.

Newnes Technical Books

The Butterworth Group

United Kingdom	**Butterworth & Co (Publishers) Ltd** London: 88 Kingsway, WC2B 6AB
Australia	**Butterworths Pty Ltd** Sydney: 586 Pacific Highway, Chatswood, NSW 2067 Also at Melbourne, Brisbane, Adelaide and Perth
Canada	**Butterworth & Co (Canada) Ltd** Toronto: 2265 Midland Avenue, Scarborough, Ontario, M1P 4S1
New Zealand	**Butterworths of New Zealand Ltd** Wellington: T & W Young Building, 77—85 Customhouse Quay, 1, CPO Box 472
South Africa	**Butterworth & Co (South Africa) (pty) Ltd** Durban: 152—154 Gale Street
USA	**Butterworth (Publishers) Inc** Boston: 10 Tower Office Park, Woburn, Mass. 01801

First published 1978 by Newnes Technical Books
a Butterworth imprint
Reprinted 1979

© Butterworth & Co (Publishers) Ltd, 1978

British Library Cataloguing in Publication Data

Johnson, W. H.
   Beginner's guide to central heating.
   1. Dwellings — Heating and ventilation
  I. Title
697'.03   TH7461   78-40192

  ISBN 0-408-00306-5

Typeset by Butterworths Litho Preparation
Department

Printed in England by Butler & Tanner Ltd,
Frome and London

Preface

In a space of less than 20 years domestic central heating in the UK changed from the obscurity of being a cumbersome and temperamental possession of the wealthy, to becoming a required fitting in all but the poorest housing. In that frenzied period men of great integrity and technical skill found themselves moving in the same direction as those who had joined in for a quick profit. And the public was usually unable to distinguish between them. Comparisons made at the time with the Klondike were appropriate enough.

That period is over, and we are safely launched upon another and gentler phase, conveniently known as the replacement market. It is of course not technically static. But it has discarded items which had small claim to survive from the last phase, and for the most part so shall we in these pages. Our intention is to give a broad picture of good current practice in its quite numerous forms, and to draw such inferences as we can for the near future of the subject.

The primary purpose of this book is to promote an understanding of central heating, so that people who intend to become owners or who have reached the time for replacement may choose with greater confidence. This in part is a matter of giving perspective, straightening the picture which has been created by unbalanced advertising pressures. We hope that those who already have central heating may learn something to their advantage about good technical and economic control of their systems. We do not see central heating installation as a subject for every amateur to tackle, and there is too much capital at risk to take chances.

The cover photograph is used by courtesy of Drayton Controls (Engineering) Ltd.

W.H.J.

Contents

1 Introduction

Before studying any subject it is necessary to establish guide lines, to define it. This is as true of central heating as anything else. For the name proves, on examination, to have been used quite loosely.

Central heating was first called 'central' for the best of reasons, that it was heating from a central source, usually a coal fired boiler. This compared with the unsystematic methods which had been in use since the end of the Roman occupation, often one open coal fire in the house, or in the last 150 years the coal fired range, which cooked as well. Larger houses with better-off occupants would generally have several open fires, quite often one in each room, which accounts for the multiple chimney stacks in buildings of the period.

The earliest domestic central heating was cumbersome, usually wholly dependent upon the meagre forces of gravity for its circulation, which made it equally dependent upon a good standard of installation so as to offer a minimum of resistance. Otherwise unexplained noises in the house were customarily and usually justly attributed to the central heating.

This state of affairs ended abruptly following work done by a research association in 1956. They showed, in the course of finding more outlets for coal, that a lot of heat could be obtained from a small boiler, with small pipes, if the water were pumped around to the radiators. This was the 'small bore' system, whose advent led to a wide range of activities. For instance pump manufacturers were forced to look into the availability and reliability of suitable models, and controls manufacturers came to life. But it was the oil industry which seized upon the system and gave it a high market rating. Only later did gas join in seriously, in due course to establish a lead as suppliers of fuel for boilers. Solid fuel, through coke and special coals, maintained a steady but low market share.

Meantime, it was not to be expected that the electricity industry would sit by and watch its competitors making hay in such a potentially vast market. They could not compete in the boiler market, and had to look elsewhere. This led to the development of the off-peak storage heater, which also offered some solution to their own generating problem, of what to do with plant necessary during the day but standing expensively idle at night. Electricity as heat energy could be stored in massive receivers overnight, so that it would leak out during the day and warm the house. The first efforts led to underfloor heating, in which concrete floors formed the heat reservoir.

Underfloor heating could rarely be applied to any but new housing, which was too limiting. So there arrived the unit heater (or storage radiator among other names) which was, and is, a rather heavy free standing unit of manageable size, incorporating a heating coil and heat absorbing material, and scientifically insulated to permit a calculated rate of heat leakage to air.

At that stage the electrical effort was just as surely random and non-central as the old coal fire had been. But it was sold quite firmly under the central heating banner, which is why the name has assumed a portmanteau quality. Only later did electricity produce the central unit, Electricaire, which is in effect a very large unit heater with ducts to other rooms and a fan to distribute warmed air.

We have often had a pedantic urge to give up using the term 'central heating' but the difficulty lies in knowing what will take its place. 'Whole house heating' is tempting but fails because it is untrue in the case of partial central heating, itself a respectable and well documented variant of central heating. To refer to 'modern heating' is too vague, too reminiscent of someone's advertising. So in the end we come back to Central Heating, which despite its imperfections is widely understood.

There is a way to classify heating systems broadly, which divides them into 'wet' and 'dry'. All the systems to be described in this book fall into one of those categories and, broad as they are, they are very important in one respect. With a wet system it may be assumed that the domestic hot water supply, i.e. hot water to taps, is taken care of, or at any rate can be covered without additional fitments. On the other hand with a dry system additional equipment is necessary to supply domestic hot water. More of that in Chapter 11.

Wet systems employ water as the heat conveying fluid. Water is extremely suitable for this duty. It is cheap, usually freely available, not outstandingly corrosive (a point to return to later), is relatively harmless if it escapes from the system, and it has a high capacity for heat. It must of course be confined within pipes, and we have already mentioned that large pipes began giving way to small pipes from 1956. More recently, small pipes have been sharing the market with very small pipes, called microbore, and this indicates how the pump industry has progressed.

Wet systems have a boiler, which puts the heat into the water. We will discuss boilers in Chapter 3. In order to take the heat out of the water at points where it is required, the most common device is the radiator. Everyone knows what a radiator is, and it is important in installation and in room furnishing to remember that it does radiate anything up to half of its total heat output. For that reason it must always be able to 'see the room', never hidden away in an alcove or behind heavy furniture. For the rest it is however a convector.

But a convector by that name is a device which gives next to no radiation, only a stream of warm air. Convectors may be natural, i.e. without fan assistance; or they may be fan convectors, which give very little warm air when not at work, but a lot when a thermostat or similar device switches on the fan. A skirting heater, which the Americans tend to call a baseboard heater, is a form of convector, and a very efficient one in terms of maintaining even room heating, horizontally and vertically. An interesting application of a wet system is that in which it serves as the heat source for a dry system. This is covered in Chapter 5.

Dry systems are, not too surprisingly, all those which do not use water as the heat conveying fluid. (It would be inaccurate to describe them as systems which use air, since nearly *all* heating systems ultimately depend upon air movement.)

A high proportion of dry systems depend upon the direct

heating of air, which is then circulated. Having pointed out
the advantages of water as a medium, let us list the advantages
of air. It is even cheaper and more readily available than water,
it is entirely non-corrosive, is not only beneficial but essential
to life (in uncontaminated form), is not a hazard if it leaks from
a conveying pipe or duct, and *it does not freeze*. Against those
qualities the only negative is that it has a low capacity for heat.

Though we will deal with warm air systems in more detail
later (Chapter 4) it may be noted here that there are two broad
types. There is the random type, in which warm air is generated
and left to find its own way around, mainly by drift, possibly
with a little assistance. This system can be quite satisfactory in
certain styles of dwelling, principally those which have a substantial vertical factor. It is not very well suited to most
bungalows, which have no noticeable vertical factor. The other
type is the ducted system, which conveys warmed air to specific
points for discharge, and then arranges by means of return air
ducts to bring most of it back for reheating.

The dry systems which do not employ air directly as the
heating medium are principally electrical. That is to say, electricity is used to heat a solid or a liquid, and in one way or
another this material passes on its heat to the surroundings. In
most cases it does so by creating a current of warmed air, which
then circulates. This includes oil filled electric radiators; storage
radiators (or unit heaters) which are mainly based upon natural
circulation but may be obtained with plain damper control or
with fan control; underfloor heating, which is no longer being
installed but deserves a mention. Storage radiators qualify for
a reduced tariff, off-peak or White Meter.

A form of electric heating in which air circulation plays no
direct part is the low temperature radiant panel. This is generally
a structural item and may be fitted to wall or ceiling. It consti-
tutes a large area source of low intensity radiation which is
invisible but can be sensed by anyone in its path.

So far we have classified central heating according to fuel
and method of operation, but there is another division — by its
scope. There are degrees of central heating and some of the
terms used are full, partial, background and selective heating.

These phrases were coined early in the development of modern central heating, and although they are not written into any legal definition or standard they are well understood in the heating trade. Let us look at them more closely.

Full central heating: a term applied to a system which will maintain all rooms at their specified temperatures simultaneously, when the outdoor temperature is $30°$F or $-1°$C.

Partial central heating: will maintain specified rooms at their specified temperatures simultaneously, when the outdoor temperature is $30°$F or $-1°$C. The specified rooms are those with heat emission apparatus installed.

Background central heating: similar to full central heating but the specified temperatures are much lower; the assumption being that other measures will be taken to boost the temperature at particular points.

Selective central heating: will maintain only a proportion of the rooms at full central heating conditions simultaneously. This assumes that the whole house is equipped with heat emitters, so that the disposition of rooms chosen to be heated may vary. This is a very sensible and economic system for a house which is, as most are, not fully occupied at almost any time.

Clearly one may ring some changes on this basic list, for example with partial background heating, or selective background heating. The important thing is that if a contractor is being employed the type of heating chosen shall be written into the contract. In all the cases mentioned there is one other factor to be detailed, and that is the specified temperatures. These are for the client to select, though some guidance is offered below. In specifying temperatures it is necessary to consider the cost of higher temperatures on the economy of running the system. The salient fact is that the cost of unnecessary heat grows out of proportion to the average cost, and the rule should be therefore to have

(1) *as much* heat (temperature) as necessary, but no more
(2) heat *when* wanted but at no other time
(3) heat *where* wanted but nowhere else.

Points 2 and 3 show the importance of good thermostatic control and on/off controls of heat emitters. Point 1 requires an assessment of reasonable temperatures, and reasonable values may be gathered from the following list.

Living rooms	$70 - 75°F$ ($21 - 24°C$)
Dining room	$65 - 70°F$ ($18 - 21°C$)
Bedrooms	$55 - 65°F$ ($12 - 18°C$) (Fashion varies widely and rapidly between cool and warm bedrooms)
Bathroom	$65°F$ ($18°C$)
Hall	$60°F$ ($16°C$)

Kitchen, if not receiving sufficient 'free' warmth from cooking etc, to be kept at approx $65°F$ ($18°C$).

Three things must be stressed about this list: first that it is for guidance only; secondly that a decision must be made about the maximum required temperature in each room, and either built into the contract or used by the client doing his own design. The third factor, which deserves to precede all others in timing, is the basis of good running economics, and the way to contribute to the national effort to conserve energy. It is, in short, to *require* less energy for a given result, by losing less. This means, insulate! Insulate to the maximum extent, knowing that money spent at an early stage on insulation will mean less outlay on heating equipment since smaller units will be needed; and it will mean savings on running cost for ever more. We shall have more to say on that important topic later. The reason for thinking of insulation even before heating is that the heat balance must be made on the *insulated* structure in order to secure all the economies in apparatus.

Before leaving the subject of preferred temperatures, and the desirability of avoiding too much warmth, we can deal with another offshoot of this. Heating 'dries the air'. A rise in temperature brings about a fall in relative humidity, so that the air feels drier. In moderation this is no more than a passing nuisance to healthy people, who acclimatise as they would if they moved to, say, the Transvaal where the prevailing humidity is very low but the climate excellent. The effect of dryness increases as the temperature rises, and it is an additional reason for keeping the temperatures down.

The word 'humidifier' is often mentioned, and it describes a device which adds water to the air. We would certainly oppose the proposition that all central heating should be accompanied by humidification, and instead we list three categories where it is usually justifiable.

(1) For people who suffer from sinus complaints. We go no further than that, and recommend that they seek medical advice by way of confirmation.

(2) Where antique furniture is involved. Furniture, unlike people, cannot adapt, and antiques might be spoiled due to shrinkage.

(3) In some cases, in older type timber framed buildings. Although the changes in temperature and humidity which take place are no more than those which occur seasonally (though these occur daily) and rarely do any harm, they do very often cause quite alarming noises to issue from the structure. It is solely to mitigate these that humidification might be considered.

Humidifiers are sold in three classes, cheap, moderate and expensive. The first is little more than a wick or similar evaporative device, though almost always made quite attractive in appearance. The prototype is a towel hanging from a radiator into a bucket of water. Commercial models of this type are suitable for most purposes, and bring about quite good results. At the extreme end of cost are models of considerable complication, extensively instrumented, electrically operated. The moderate range falls somewhere between those two extremes.

So much for shortage, or apparent shortage, of moisture in the air. But what about the times when there seems to be too much, when it forms as condensation, stains wall paper, ruins paint and curtains, mists windows and turns clothes mouldy in wardrobes?. There is no magic machine which will *reduce* the relative humidity irrespective of temperature, but there are two ways to avoid condensation. One is to prevent wall surfaces,

in particular outdoor wall surfaces, and windows from becoming cold enough to cause it. The other is to ventilate, to
remove enough of the moisture laden air as it forms to keep
the concentration within manageable limits. These matters we
will deal with when we come to Insulation.

There is no such thing as a best heating system. What is best
for any one situation depends upon such things as what it is
expected to do, where it has to go, and what may not be
available (such as a ready fuel supply).

One such, and a common feature, concerns the space
required compared to the space available. An outstanding
example of this is a warm air system with full ducting. Ducts
average about 250 x 200 mm or 10 x 8 in. Very few houses
can tolerate apparatus of such a size running down a corridor
or across a ceiling. For that reason, a full warm air duct system
is always regarded as something to be built into a new house,
not added to an existing one. At construction stage ducting
can be built into the structure and 'lost'. Warm air systems put
into existing property tend to use stub ducting, i.e. short runs
leading only into rooms adjacent to the heat unit. Underfloor
electric heating was of course associated with new building,
since it was almost wholly incorporated in concrete floors. Its
successor, the storage radiator, is sometimes described by
owners of rather small rooms as bulky or cumbersome. Some
manufacturers have gone as far as the natural limitations of
the subject will allow in reducing the depth of projection into
the room. In such cases an unbiased observer might say that
the unit takes no more space than would a small bookcase, and
it has a safe surface temperature. It can therefore be assimilated
into the room.

Still considering forms of electric heating, radiant wall and
ceiling panels take no noticeable space, but like warm air
ducting must be thought of as structural items (though the
upheaval in fitting them into existing premises is often within
reason).

Let us examine the space and facilities demanded by wet
systems. Until quite recently the boiler would stand alongside
the refrigerator or the washing machine and look very like either.

Or, still the same size, it might be situated under the stairs, or in a spare room. The criterion was often where it could get access to a flue, either of the conventional or the balanced type (see Chapter 3). Nowadays, a boiler may be hidden behind the gas fire, so taking no useful room at all; or hung on the kitchen wall like a moderately sized cupboard.

Radiators cannot be reduced in size since surface area is fundamental to their performance. Within that context however they have been streamlined, brought as near to the wall as possible without impairing their efficiency, and have had projections smoothed away. The old pattern column radiator is scarcely available now, though it, and the double and treble panel versions of the panel radiator, continue to supply the answer to that other problem, of where there is enough space for projection but too little wall area for the required size of single panel. Radiator manufacturers do their best to meet problems of wall space by offering a choice of heights of radiator. For example the heat emission from a radiator measuring 4 x 1 is just about the same as that from a panel 2 x 2. Shortage of wall space may be overcome in two other ways. The skirting heater increases the projection of a normal skirting board but otherwise uses no wall space at all, and the fan convector may be assumed to occupy a wall area only one sixth or less of the area of a panel radiator of comparable output.

In an extreme case, where the wall width was only two feet, but the height ample, we have known a long radiator of limited height to be fitted vertically instead of horizontally in order to get the required area into a room. It shows what can be done, but for technical reasons this ploy is not highly recommended.

The problem of fitting an appropriate type of system to a given set of conditions has other aspects besides space. Some may be overcome. For instance, no gas main within miles? Then there is bottled gas. Some are not overcome so easily, if at all. Take for instance a flat in a multistorey block. It will almost certainly have no conventional flue. Whether a balanced flue would be allowed to pass through the building wall must be determined by reference to the conditions of sale, or lease, or let. Such a flat would quite certainly not be able to accommodate an oil fuel storage tank, and that limitation is much more widely applicable, covering all premises which have no garden or very little garden.

When we come to the point of deciding which form of central heating to choose, the choice is not so wide and bewildering as might at first appear. Let us begin by supposing that the whole of available heating stretches from A to Z. Then, by the time we have eliminated those sections of the market which cannot be accommodated by us, the market has shrunk to, say, A to P. Then, we find that we do not much care for, it may be electricity, or gas. That would reduce the choice to A to J. Within the new short range we must then be careful to see that what we look at has earned the Approval of the appropriate authority: SFAS for solid fuel, British Gas for gas, DOBETA for oil, either BEAMA or the Electricity Council for electricity, MARC for radiators. All goods which are sold in appointed places, such as gas and electricity showrooms, can be taken for granted. Nowhere else should one fail to enquire about Approval. There are, it is true, cheapjack wares about, but decreasingly so. More to the point are the imported items which are reaching us. There are of course many excellent items being imported, but an item is usually designed for the conditions it will meet in its home country, which are not necessarily those of the UK. It could therefore be successful at home but not here. Any item which is both reputable and suited to UK conditions will have been submitted for UK approval (just as UK items must be tested by countries importing them). You will see therefore that whether an item is British made or imported, you need to satisfy yourself that it has Approval. More than that, if you place your heating contract in the hands of an installer, make sure that he chooses and uses Approved items on your behalf.

Practical points to emerge from this chapter are:
  (1) Not all types of heating system can be accommodated in every home.
  (2) There are two broad types of system – wet and dry.
  (3) The range of choice within 'wet' and 'dry' allows for at least one example of each to be suited to any home.

(4) Before setting out to acquire a heating system (or even a simple gas fire) the householder should be quite clear about the standard of heating he is seeking, e.g. whether full, partial or background; and the maximum temperatures to be maintained.

(5) When buying apparatus make sure that it complies with an official Approval scheme wherever appropriate.

2 The Wet System

For far too long central heating has been known by the fuel which supplies the heat energy. We have for instance gas central heating, or oil central heating. This is entirely due to the efforts of advertisers, of whom the biggest and most persistent spenders have been the fuel interests.

The truth of the matter is that the important aspect of a system is the heat emission side. Suppose you have radiators installed, and a wet system, with a gas boiler. It is a couple of hours work to remove the gas boiler and fit instead a coke or oil boiler. But to take out the radiators and fit instead ducting or some other type of system is a major operation. Equally, substituting a boiler makes no difference at all to the heating, whereas any departure from the radiator system will cause some noticeable change, probably introduce a new method of control for comfort. That is why we are going to consider wet systems before going on to boilers.

In spite of a few half hearted attempts to make changes, the wetness of wet systems is water. Partly with car practice in mind it is possible in a closed circulation system (q.v.) to introduce an antifreeze solution. Another additive is aimed at reducing internal corrosion, and this is vehemently defended by its suppliers and supporters but by no means universally accepted.

The operation of a wet system is entirely dependent upon circulation. Water is heated in the boiler, is piped to the heat emitters (radiators etc.) and returns, partly cooled, to be reheated. In earlier days, and to a very limited extent nowadays, heating circuits operated by gravity. A warmed column of water is lighter than a cold column, and so the one rises and the other falls. A system based upon this natural movement has to have a substantial vertical component. The rise and the fall must be fully established in order that any sideways movement, into radiators etc., can take place with continuing circulation and so with renewal of heat introduction.

The principle of gravity circulation is still the one most used on the domestic hot water side of combined systems. The pipes, known as the primaries, flow to and return from the cylinder, which must be at a higher vertical level than the boiler.

But heating circuits cannot rely upon finding a high vertical to horizontal ratio, except perhaps in a lighthouse. Most gravity inspired heating circuits were either sluggish or failed to circulate. Those that worked were characterised by a wide difference between flow and return temperatures (due to slowness of travel) so that radiators became progressively cooler along the line. The gravity circuit has passed, therefore, without regret. Its successor is the pump assisted, or pumped, circuit, which began seriously with what became known very quickly as Small Bore. Small bore employs pipes mainly in the range 15 or 22 mm, occasionally 28 mm (which used to be ½, ¾ and 1 inch). Gravity systems could never fall below 1 inch pipe.

The reduced diameter, the extra mobility contributed by the use of copper instead of iron, the design freedom due to pumping, have all led to small bore being widely and readily applicable. It is no more difficult to make work in a single storey barrack style building than in a conventional semi. It can of course be made to misbehave, for example by having too much pump power, or by using pipes inadequate for the designed load at any point in the system. But those are matters to be sorted out during design, as we shall be showing in Chapter 12.

Microbore

After twenty years of adherence to small bore we have more recently come to microbore, in which the pipework is almost entirely smaller than 15 mm or ½ in, being more of the order 10 mm. This material can be installed in the same fashion as medium weight electric cable, which makes it a good deal easier, quicker and therefore cheaper to handle than small bore tube. The material used is copper, though nylon is on offer. The latter must be viewed with some reservation, since its only real advantage is in first cost.

The pump to be used with microbore develops more pressure than a small bore pump, since the system still has to carry the same weight of water in unit time. Microbore circuits have another notable difference, compared with what has gone before. In other sizes of circuit the pipe goes from one appliance to the next, giving up as much as each requires. In microbore circuits there is no reserve for such tributary treatment. The total amount of heat for the system, as hot water, is led to a central distributing point called a manifold, by a flow pipe of standard (not microbore) size. The manifold (*Figure 2.1*) should

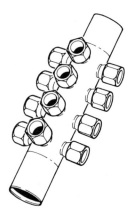

Fig. 2.1. A typical micro-bore manifold. This is fitted under the floor or in a cupboard and distributes pipes to the various radiators

have as many connecting points as there are heat emission appliances. From this central distribution point a pair of microbore pipes is led to each appliance, by the shortest route, and connected to the inlet and outlet. The microbore pipe connected to the appliance outlet is then connected, back at the centre, to another (return) manifold (*Figure 2.2*), which in turn is connected to the return pipe to the boiler.

In spite of the ease of pipe running and other advantages of microbore, it is not destined to take over the market, though it will achieve a share. To give only one reason: a gravity circulation system could be brought to a standstill by almost anything — a dip in a supposedly horizontal pipe, or an air bubble; a clean small bore system will fight on through most disturbances, short of having scale or similar foreign matter choke up

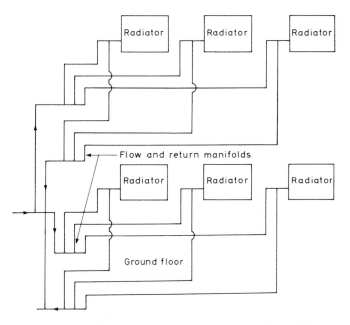

Fig. 2.2. Typical microbore system showing the use of manifolds

the pump; the cross section of microbore pipe is so small that it can be blocked by quite small pieces of foreign matter. We mention this because there is always a tendency to imagine that the latest of anything is about to become the only one. In the end anything of value assumes a reasonably constant share of the market after a settling in period.

Small bore

Small bore is likely to dominate the wet system market for quite a while. It is split into two categories, single pipe and two pipe systems, which we will examine. At present the balance lies in favour of the single pipe system, but only on account of first cost. The two-pipe system, which is technically a much better job, uses almost twice as much pipe. Perhaps we can do something, in the next few pages, to swing the balance in favour of the better system, by pointing out how the results of recent investigations can lead to overall savings, thus bringing the cost of two-pipe to the level once expected from single-pipe. But first, to see what single and two-pipe systems are. The differences are readily apparent from *Figures 2.3* and *2.4*.

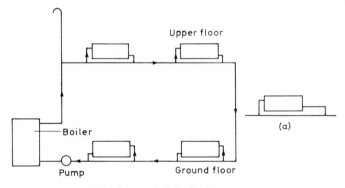

Fig. 2.3. A typical single-pipe system

Taking *Figure 2.3* first, showing a single-pipe system, the flow pipe will be seen to leave the boiler and travel along the upper storey floor. Each radiator is connected so as to take water from, and return water to, this pipe. The process is repeated on the ground floor, until eventually the single pipe, now known as the return, enters the boiler.

At any one radiator there are two important things to note. First, any water which passes through the radiator gives up heat and so drops in temperature. Consequently the water in the flow pipe, just after the point at which the radiator returns to it, will be cooler than before due to the cooling effect of the water from the radiator. But that same water is of course the water which enters the next radiator, and so there is a progressive cooling of the heating medium, as the flowing water is called, along the circuit. This same thing occurred, of course, with gravity systems, and the only benefit of the new system is that thanks to pumping the temperature drop is roughly halved.

The second thing which should be noted is that, while the system is pump circulated, the circulation *within each radiator* depends upon natural forces. The force causing flow to start within a radiator is the very small difference in pressure in the flow pipe due to the frictional resistance of the length of pipe between radiator inlet and outlet. That is why, in some cases of reluctant circulation, a cure has been made by transferring the radiator outlet connection to a point further downstream on the circulation pipe (*Figure 2.3a*). Reluctance, it may be added, is rarely so fundamental. No radiator should be bought without an air valve key, unless an automatic air vent is fitted. That key should be used whenever sluggish behaviour is noticed, and at regular intervals if there seems to be a tendency to accumulate air.

The two-pipe system is shown in *Figure 2.4*. It can be seen that the flow pipe goes only as far as the inlet to the last

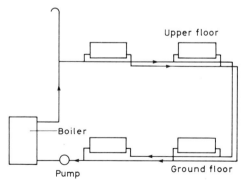

Fig. 2.4. A typical two-pipe system

radiator in the circuit, and that the return pipe begins at the
outlet of the first radiator. Thus, in the *Figure 2.3* scheme, a
single pipe acts for most of its length as both flow and return.
But in the two-pipe system the flow and return functions are
kept apart, the return coping only with that water which is
waste, or spent, from the heating process.

It will be obvious from this that, subject only to a small
practical loss, each radiator receives its water at the same temperature, which is the predetermined boiler temperature. This
in turn means that (a) the heating system has more chance of
being fully effective, and (b) it can actually save money on
radiators, since to get a given heat output from a radiator
operating at a lower temperature (as those later in the single-
pipe circuit would do) calls for a larger radiator.

The second thing to be noticed about *Figure 2.4* is that each
radiator is connected, inlet to flow, outlet to return. There is
thus a wide differential pressure causing flow though the
radiator, and the manner of flow is much nearer to being
positively pumped. This means that the velocity of water flow
through the radiator, hence the rate of warming and the
response to change, are all more rapid.

A further advantage to come from the two-pipe system and
its near-positive water flow is that we need no longer rely upon
gravity circulation within the radiator, as was the case with a
single-pipe system. Consequently we can get away from the
'top and bottom opposite ends' method of connection shown
in *Figure 2.3*, and use any of the much tidier arrangements.
Typical is that shown in *Figure 2.4*, of both connections at
the bottom; or a later version which employs only one radiator
tapping into which both connections enter.

Indirect Systems

All systems, whether single or two-pipe or microbore, are
capable of being installed in two ways – direct and indirect.
They should certainly not be installed as direct systems, but
we must at least consider the possibility in order to see why not.

A direct system is the basic one, in which water passes from boiler to pipework and back to boiler. It goes, too, usually through a cylinder, to hot taps. So every time hot water is drawn off, more water enters the system and passes through the boiler. Raw water is a mixture of many things, depending upon its source, but most of the ingredients can be harmful to a heating system, in or after going through a boiler. Hard waters will deposit scale in and beyond the boiler; reducing waterways, coating heating surfaces so that they must be forced to higher temperatures and in the end break from heat fatigue, or burn out. Many soft waters act in quite another way. Their reaction being acid, they attack metal surfaces, and bring about corrosion. This usually destroys a system more quickly than the scale build-up in hard water systems.

Direct systems give a choice, therefore, between the devil and the deep blue sea, and the safest course is to have no dealings with them. The way out is by an indirect system, in which water drawn from taps does not pass through the boiler, but gets heated indirectly. The key apparatus is an indirect cylinder. In appearance like any other cylinder, it contains a heat exchanger, in the form of a coil or other suitable type. Boiler water passes through the heat exchanger, and the domestic hot water for taps passes around the outside of the heat exchanger.

Thus it is that in a leak-free indirect system the primary circulation, which is the water passing through the boiler, heating system and heating coil in the cylinder, never changes. It is the same water going round all the time. If we suppose a system to contain 45 litres of water, and the water in raw condition to have in it 100 parts per million of scale forming or acid ingredient, the total weight of that ingredient to enter the system is 4.5 grams, which is negligible. Raw water must of course enter the cylinder on the secondary side, i.e. outside the heat exchanger, with its scaling or corrosive potentiality unchecked. But this is nowhere near as detrimental because

(1) under the less drastic heating conditions it does not behave so badly;

(2) it is easier and cheaper to clean or renew any apparatus on the secondary side, if this should ever become necessary; and

(3) it may be avoided entirely, since water treatment, most often softening, can be applied to the supply to the cylinder secondary side.

Water Treatment

Water in this context is rarely treated for acidity, mainly because acidity under the gentler conditions is rarely a great problem. Water softening however is often practised for a number of reasons, if nothing more than the supposition that hard water is bad for washing hair. There are really two levels of softening treatment. There is that which employs a softener, a piece of apparatus interposed in the line. Permutit is a long established example of this. The other method consists in treating the water in the feed/storage cistern, by suspending in it a container of crystals which dissolve slowly in the water and inhibit any scale material from depositing in crystalline form. Calgon or Micromet is obtainable, in suitable containers, from builder's merchants.

Let us pause and think what we know so far about a preferred type of system. It is very likely small bore, and a two-pipe system. It is indirect, perhaps with water softening applied to the secondary water supply. There is something else, which has become possible only in quite recent times. It may be open or sealed. The system which we all know in the UK (*Figure 2.5*), which has one or sometimes two cisterns in the roof, is an open system. It is precisely at these cisterns that it is open – to atmosphere. This means that in the ordinary way it cannot build up more pressure than that which is due to the 'head' of water measured up to the cistern level. If there were a surge of pressure it would vent off safely at the cistern.

The open type of system, though it proves satisfactory to most people, has been an irritant to some technical workers for a long time, partly because its supporters have claimed for it blessings which, apparently, the rest of the world cannot see. For it is almost exclusive to us. Not least of the complaints is that it forces us to store a large amount of water in our attics, and, some might say, makes us nationally dependent upon the ball valve.

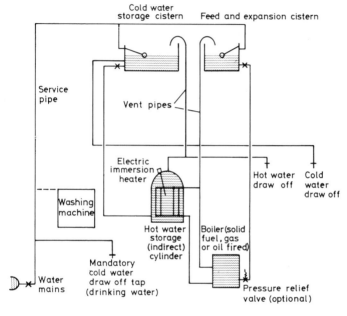

Fig. 2.5. Typical British domestic water installation

The alternative to an open system is a closed or sealed system, and this may be accompanied by either cistern fed or mains fed domestic hot water. The differences are shown in *Figure 2.6* and *2.7*.

It must first be understood that sealed systems are capable of building up pressures beyond those of an open system. This means that most of the apparatus — boilers, cylinders etc, supplied for open systems may be neither suitable nor safe. It means too that the general standard, even of making joints, must be higher; and that certain extra items must be fitted, which low pressure systems do not require. Thus in cost terms any economies due to saving a cistern and pipework are wiped out.

If we couple that with a warning, that sealed systems are not yet through the wood in terms of total acceptance by authorities; that high pressure appliances are not yet easy to come by; that the installation work is not for amateurs and not yet for a lot of professionals; we hope that readers will accept that in ordinary circumstances there is a lot of useful life left in open systems.

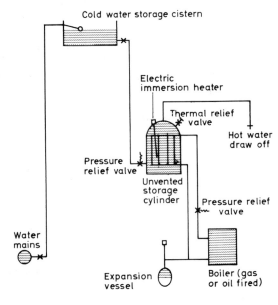

Fig. 2.6. Cistern fed unvented domestic hot water system with a sealed primary circuit. At the present time there are no bylaw requirements for the provision of a pressure relief valve, a thermal relief valve or an expansion vessel. The precautions shown are those recommended by the P.H.W.S. committee

A *sealed primary system* is one in which only the primary system is sealed, the secondary continuing to be supplied from a cistern. This is shown diagrammatically in *Figure 2.6* and includes the all-important expansion vessel. This vessel, in which a rubber diaphragm separates the water in the system from a cushion of air or nitrogen, takes the place of the expansion cistern in absorbing any expansion taking place during heating of the water.

The advantages claimed for a sealed primary system are:

(1) No cold feed/expansion cistern and consequently no freezing problems.

(2) No need for the head (vertical elevation) of a header tank, and the whole system can be accommodated on one floor level. Note that this is advantageous *only* in a system solely for heating. If domestic hot water is added, a loft cistern *is* required.

(3) It follows from (2) that if convenient the entire system may be accommodated in the loft to save useful space, subject to the same condition as (2).

(4) Sealing eliminates oxygen pick-up, thus reducing one source of corrosion.

(5) A sealed system can operate at higher temperatures since the boiling point of water is raised according to the amount of pressure imposed upon the system (determined by the control instruments). This means that smaller heat emitters may be used for a given duty. It will eliminate any tendency to boiling noises which afflict some installations.

It is a fact, but not an advantage, that if higher flow temperatures are used, conventional radiators cannot be used since they will develop unsafe surface temperatures. Convectors must be used instead. A distinct disadvantage of these systems is that, at the present time, they may not be connected to the mains water supply. They must be inspected regularly to see whether there is a loss of water, and if necessary topped up by manual means.

A *mains fed sealed system* as shown in *Figure 2.7* is fully pressurised, in that the domestic hot water supply is taken

direct from the mains. This, it is fair to predict, will be the
common pattern of the future, and not only on technical
grounds. But to look at the technical factors first, this system
begins by claiming those advantages listed for the sealed primary

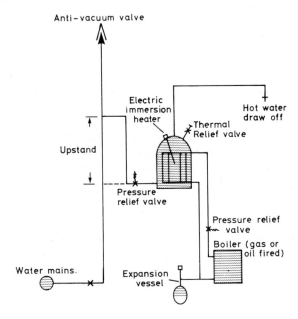

*Fig. 2.7. Mains fed unvented domestic hot water storage
system with a sealed primary circuit. Precautions shown are
those recommended by the P.H.W.S. committee. Some of
these are mandatory in continental Europe. The combination
of anti-vacuum valve and upstand is required to prevent the
water heater from draining down; similar protection can be
afforded in other ways*

system. It goes on to claim, for the pressurised secondary or
domestic hot water side, that:

(1) it overcomes complaints about low pressure systems
not giving sufficient flow or pressure at appliances. The

outstanding example of this is the shower, which in the UK is on average so much worse than continental practice;

(2) it makes possible a high performance total water installation on one floor, without need of gaining height (as in a loft);

(3) it clears the way for much smaller and neater water terminal fittings, taps and mixers, and for interconnections of smaller pipe diameter.

A. Cistern fed unvented hot water storage and sealed primary circuits — see fig 2·6

B. Instantaneous water heaters

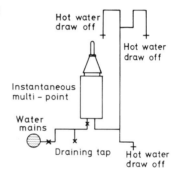

C. Water jacketed tube heaters

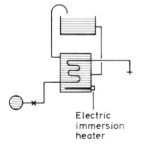

D. Mains fed storage heaters (each up to 68 litres)

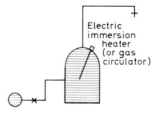

Fig. 2.8. Examples of unvented hot water apparatus which comply with the model water bylaws

The third point is an important one, with economic as well as technical issues involved. If UK water fittings can conform to continental standards in size and performance, which is to say an E.E.C. standard, we can use imported fittings and our manufacturers can make for continental markets without the extra cost of making 'specials'.

The sealed primary system is being installed, to a limited extent, and if a gas boiler is involved the gas industry is able to set out the limiting conditions for a safe installation.

At the time of writing a lot of high level activity is taking place in an effort to spell out the basic requirements of a wholly pressurised system. This involves the National Water Council and the regional Water Authorities, the Department of the Environment, and interested bodies such as the Institute of Plumbing. The impetus and the ingenuity will come from industry, who will design and make apparatus capable of performing right up to − and sometimes beyond − the limits to be laid down by the legislative bodies.

Pumps

The heart of a modern heating system, the pump, is still referred to at times as an accelerator or a circulator. More important to the user is the improvement in its reliability. In the early days of small bore the pump was the most temperamental item in view, if not actually breaking down then noisy. Nowadays all that a pump needs is to be protected from too much foreign matter in the circulating water and it will generally go on working indefinitely. We will not attempt to explain the internal arrangements of a pump since this is rarely relevant to the user or even the installer. Pumps may be 'in line' fixed, and consequently supported by the pipework and brackets; or they may stand on the floor. The former arrangement is by far the most common, particularly since pumps have become more compact. They do require adequate support, usually achieved by pipe bracketing but it is not necessary to use flexible connections as was being advised little more than 10 years ago! Makers' recommendations should be followed. For example some makers require that their pumps shall run with the spindle horizontal, which means fitting the pump in a vertical pipe line.

Many modern pumps have variable output adjustment, so that the pump, up to the limit of its capability, may be adjusted to suit the requirements of the installation. Broadly speaking a pump must be able to achieve the design requirements, which may be measured as causing a drop of 10 degC or 20 degF across the circuit. It must not do so much that it tends to push water right out of the circuit through the vent pipe. This condition, known as 'pumping over', encourages the water to dissolve air before it returns to the system. In that state it is responsible for massive internal corrosion of the system.

A pump which does not have adjustment built in may be suitably controlled by regulating the gate valve on the *outlet*. All pumps should be fitted with gate valves on inlet and outlet, so that the pump may be serviced without draining down the system. If an outlet gate valve is used as a regulator it is very helpful to attach a label to it, stating the amount of closure applied, e.g. ¾ turn, so that if shut for any reason the gate valve may be reset without a trial-and-error period.

Other forms of pump now available are the two-speed pump and the twin pumps in parallel. The first may have some application as a slow speed runner when the system is, relatively speaking, idling, as in late spring. Otherwise, the claim that it is suited to both large and small systems is not likely to benefit the householder if he has to pay for such versatility but never needs it. The other, the twin pumps, may suit those of a nervous disposition who believe in being prepared. It is however a development which would have been much more welcome ten or more years ago.

An important, indeed fundamental decision in circuit design is where to put the pump in relation to the boiler; and since the purpose of this concern is to relate the pump correctly to the vent and the cold feed, the decision really involves all three. Let us first dispose of one theory which was devised more as a money saver than on technical grounds, namely that it is satisfactory to combine the vent and the cold feed. It is not satisfactory and is bad practice.

The pump is more often fitted in the return than in the flow. In a small bore unit which comes fully equipped including a piped-in pump, the pump is almost always connected to the return. The return-fitted pump suits the average installation, in which the feed/expansion cistern is in the loft, and the highest point of the heating circuit is no more than about 0.6 metres above the bedroom floor level. There is then a distance between the cistern and the highest point in the circuit greater than the head of the pump. This creates a positive head of water sufficient to counteract the 'suction' effect which the

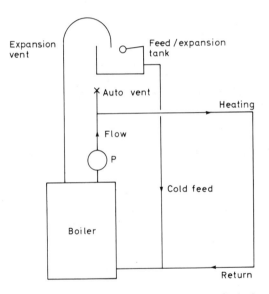

Fig. 2.9. Pump position where head of water is limited, for heating only

pump has on a considerable length of circuit, which would otherwise encourage air infiltration. If the vertical distance between the water level in the cistern and the highest point of the circuit is small, and less than the pump head, then the pump must be fitted in the flow pipe (*Figure 2.9*).

The cold feed and expansion or vent pipes must then be connected both on the same side of the pump. A useful arrangement with pump in the flow is shown in *Figure 2.10*. The air vent pipe is taken off the second boiler flow tapping and is an extension of the hot water flow primary. Like the

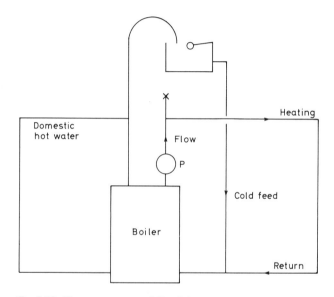

Fig. 2.10. The arrangement of Fig. 2.9 with provision for domestic hot water

cold water feed it is effectively on the pump suction. *Figure 2.11* shows a comparable system but with the pump in the return. This is an appropriate point at which to mention another detail of design. In both *Figures 2.10* and *2.11* it will be noticed that the heating return enters the boiler opposite to the hot water return pipe. (In boilers which do not have tappings opposite this does not apply.) We have already mentioned that the force causing gravity circulation is a very small one. It does not take a lot to neutralise or even to reverse it.

30

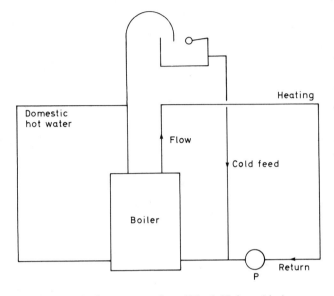

Fig. 2.11. A similar system to that of Fig. 2.10, but with the pump in the return pipe

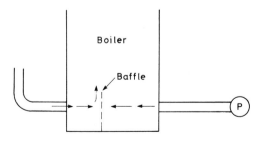

Fig. 2.12. Position of baffle in boiler to separate flows of heating and hot water circuits

The situation, when the pump is running, is illustrated in *Figure 2.12* and this can in most cases promote reversed circulation in the hot water primaries for at least as long as the pump works. If the pump works for, say 10 minutes in every 20 or 30 minutes, the hot water circuit is unlikely to settle down to work. In some boilers this is recognised, and a baffle as shown in *Figure 2.12* is put in to keep the two streams apart.

Where no such arrangement exists the solution is to make both returns enter on the same side of the boiler, which means using a common pipe. The scheme shown in *Figure 2.13a* would

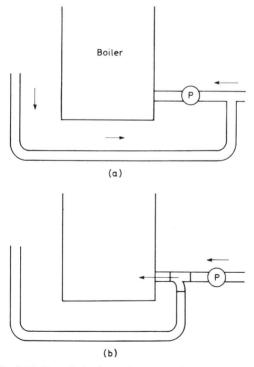

Fig. 2.13. Natural circulation is prevented in (a) when the pump is not in use, but possible with the arrangement shown in (b)

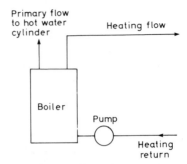

Fig. 2.14. Boiler and pump connections for a small-bore heating system

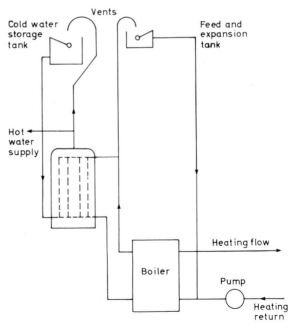

Fig. 2.15. Main features of a typical combined small-bore heating and hot water supply system

be excellent in lending pump suction to the hot water primary and so giving it a useful tonic, but it would effectively stop any circulation when the pump is off, which is all the summer. Consequently we come to arrangement *Figure 2.13b*. By the use of a swept or pitcher tee the flow of water from the pump is used to induce a flow in the other pipe, and if the pump is not running a natural circulation can carry on.

We must now come to consider the appliances which are placed in a wet system circuit as heat emitters, and we start with the most common one, the radiator.

Radiators

Most radiators nowadays are of the panel type. This means that they are almost flat but are grooved or dimpled or in some way wrinkled in order to increase the surface area within the total size which is called the 'picture frame' area (*Figure 2.16*). On this surface area their output depends.

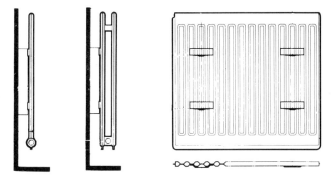

Fig. 2.16. Typical panel radiator which may be single or double panels (by courtesy of Thorn Heating Ltd.)

The other traditional design, called the column radiator, is rarely good looking, being designed for high output in a given space, and often known as the hospital radiator. Although not likely to be considered now for domestic use, it is if met the only radiator normally equipped with feet for floor standing. (This excludes some oil filled panel radiators for electric heating, which are fitted with feet.)

An imported style of radiator which is different in design is that which has a plain front face, e.g. of aluminium, with a heating coil closely soldered or brazed to the back. Radiators of this type offer a high resistance to flow and *must* be supplied through a two-pipe system or microbore, not a single-pipe system.

If other types and designs crop up from time to time they must be examined on their merits and it is up to the suppliers to say what those merits are — and to give some proof of what they claim, in particular the nominal heat output upon which all calculations must depend. The safest procedure is to insist upon buying only what has received MARC approval. (MARC is the Manufacturers' Association of Radiators and Convectors).

All radiators, being vessels which are capable of collecting air, should be equipped with an air release tapping, to be fitted with an air cock or automatic air vent. All radiators should be supplied with the proper means of support, which usually means wall brackets.

Radiators may be finished in enamel or left in primer to be finished by the customer. In the latter case, and for subsequent redecoration, it makes very little difference what type or colour of paint is used, with the one exception that so-called metallic paints, such as aluminium and bronze, must not be used since they reduce the heat output substantially. It is worth noting here that after repainting three or four times the thickness of paint film will become a barrier to heat movement and the paint should then be stripped before repainting again.

The output of a radiator is one aspect examined by MARC. It is measured as the heat emission from the panel when there exists a temperature difference, between the water inside the panel and the air just outside it, of 55 degC or 100 degF. Obviously the emission will be higher if the differential is greater, less if it is lower. Perhaps less obviously, a radiator will give of its best when first coming to work in a cold room. As the room warms the output rate will slow down in proportion, this occurring naturally, without the use of instruments or controllers. Though we must rely upon MARC for a definitive output it is often convenient to get a rough idea of how big a radiator is going to be needed before the design is finalised.

In rough figures, then, a single panel radiator will give 0.6 kW/m^2 /55degC or 190 Btu/ft^2 /100 degF: while a double panel will give 0.5 and 160 respectively.

Using those figures to find out just how big a radiator is in wall or picture frame area, take the dimension figure, square metres or feet and remember that all radiators have two sides. So, suppose a radiator surface is given as $1m^2$ ($10ft^2$), the side you see is only $0.5m^2$ ($5ft^2$). For approximation this may well suffice, but it will be a trifle on the high side. The surface area of a panel, you will recall, is artificially increased by the use of profiling. To get a nearer idea of the picture frame area you could knock 10% off the nominal figure, in this case giving $0.45m^2$ (4.5 ft^2). Steel radiators are recommended for use with indirect systems only.

Where to Put the Radiator

We must start any such consideration by knowing how a radiator — any radiator, whether water or oil filled — works. It delivers heat as natural convection, a rising warm air current, and as radiation. Since the radiation content can be up to half the total, its importance cannot be overlooked. This leads to the first basic conclusion, that wherever the radiator is situated it must be able to 'see' the room. That is to say that it must not be hidden, either permanently behind a sideboard or similar, or effectively by having an easy chair usually pushed back against it. As well as shielding the radiant warmth from the room such treatment will result in overheating the adjacent furniture, with rapid deterioration of textiles and plastics.

We must then think about the radiation from the back side of the radiator. Radiators are always fitted to walls or panels, and usually by brackets supplied. Tests have shown that there is an appreciable loss of heat from the room, through the wall, for this reason. Further tests, interposing insulating slabs and reflective insulation between radiator and wall were not encouraging. The first showed no real change, while the second reduced heat loss but also reduced radiator output.

The logical conclusion is that the radiator would be best placed in the middle of the room. In practical terms it means that it should be fitted to an inside wall, and if this is done overall heat savings of the order of $5 - 10\%$ can be expected. In addition there will usually be a significant saving in pipe runs. This location is in obvious conflict with what we have come to accept, that radiators should be fitted under windows. We must remember however that there was just one reason for that, namely to counter descending downdraughts off the cold glass, with reduction of condensation. That remains a valid reason, except that it ceases to apply where double glazing is fitted. The under-window position has always introduced contention with ladies whose attachment to their full length curtains has assumed greater importance than the technical refinements of heating practice.

Having established that there is an equal case for fitting radiators on inside walls or under windows, we come in the former case to ask where. Within broad limits this is not import-ant. It calls for a small amount of imagination, a visualisation of how the warm air will flow as it rises off the radiator. For example, given only one radiator in a room which is long and narrow one would not fit the radiator at one end, but roughly central – unless occupation were to be confined to one end. The situation chosen for the radiator must allow not only for the radiator to 'see' but for a free air flow to have access to it. Radiators must not be boxed in, for appearance or any other reason, unless it is fully appreciated that this can result in a serious fall in heat output.

A more widespread move to take radiators away from the window position will draw new attention to the radiator shelf. This device has a strictly limited purpose. It prevents or miti-gates wall staining above the radiator, which is justification enough. Its effect upon radiator output is usually to bring about a slight decrease, and care must be taken to fit it in the right relative position, not for instance sitting down tightly upon the radiator.

The conclusions about situation reached for radiators do not apply to convectors, fan convectors or unidirectional radiators (such as electric fires). All appliances which are predominantly convectors will have positions which suit their warm air output whether natural or forced. It is a pity that warm air cannot be coloured, since this would greatly assist the imagination when choosing the best sites.

Connections to Radiators

Having examined the general form of the system, and expressed a preference for the two-pipe version, we come to the detail of radiator connection to any type of system.

The wall brackets must be securely fastened to the wall, in positions so that the radiator is suspended true and level. The final connections should be made between circulating pipes and radiator so as to impose no strain upon the brackets or the radiator.

Each radiator should have two valves fitted. It is bad practice to skimp in this respect, for even removing a radiator to paint it will involve the cost or inconvenience of draining down the whole system. One of the valves should be of the lockshield pattern, i.e. to be turned only with a special key. This valve is used for balancing when commissioning the system, as we shall explain later. The amount of opening should be recorded in case it is closed for any reason.

It is a good rule to fit the lockshield valve on the outlet end, the other valve for on/off control being on the inlet. The only time that order becomes obligatory is when the control valve is of the thermostatic type, since this type is almost always made for unidirectional flow. The on/off control is there to be used. It is the means of achieving warmth only when and where needed. The use of thermostatic radiator valves is explained in Chapter 8.. In general the angle pattern valve is preferable to the straight pattern. Not only does it make for a neater job, with less pipe showing, but the radiator may be removed from service without having to spring the pipes apart.

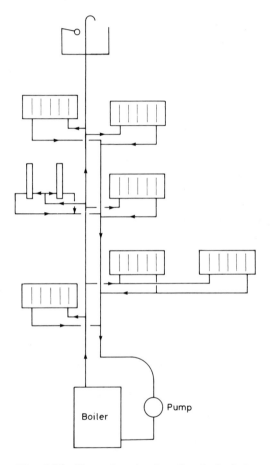

Fig. 2.17. Illustrating the benefit obtained by running flow and return pipes by the most direct route. Only the radiator circuits are shown

The Close-coupled System

The growing practice of double glazing has greatly reduced the significance of radiators under windows. It opens the way for considerable reduction in installation cost, particularly in premises which tend to be long and narrow, in either a vertical or a horizontal direction. As *Figure 2.17* shows, the benefit is achieved by running the flow and return out and in by the most direct route, taking off connections to radiators by the shortest pipe runs. In some cases radiators may be fitted back to back, the ultimate in pipe economy. A system may combine this type of run with longer runs made inevitable by the house plan. *Figure 2.17* shows a vertical system, but the principle applies equally to a horizontal run. Also, the diagram shows a two-pipe system but it can apply to a single pipe system.

Skirting Heaters

The radiator, however much it is styled, is an alien object in any home. If it were not wanted for heating no one would have it. But everyone has skirting boards, and heaters which closely resemble skirting boards must be widely acceptable. That is just one of the advantages of skirting heating. Others are that it does not take up wall space, that it encourages a better vertical temperature gradient (difference between temperature at floor and ceiling level). The reason that we do not see more of it in the UK is that it costs more than radiators, but another answer might be that we sometimes run out of wall. Walls are often used to back sideboards, bookcases, cupboards, thus effectively making those skirtings unusable.

For a rough check, once the heat requirement of the room has been calculated, an average skirting heater will give 450 watt/metre or Btu/ft length of heater. This is based upon a mean temperature difference of 55 degC or 100 degF. In marginal cases a search of the market might show that one make offers significantly more output per unit length.

The skirting heater is, in effect, a tube whose heat emission characteristic has been greatly increased by a number of fins

brazed to it. This is encased in a panel structure fitted with a
damper (*Figure 2.18*). The damper controls the amount of air
able to flow over the fins, hence the heat output of the unit. A
closed damper will generally bring the output down to about
30% of the rated value. A more satisfactory method of construction is that which employs a bypass pipe with valves on the
finned tube, so that when not wanted in circuit it may be
isolated.

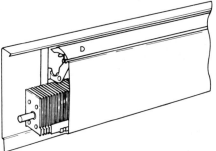

Fig. 2.18. Typical skirting heater in section.
D is the damper

These notes, on skirting heaters and similar apparatus, are
necessarily general in character, and aimed to explain the nature
of the items. Each manufacturer introduces his special feature,
which may add something to ease of operation, or appearance,
or ease of maintenance. Once the reader is aware of the broad
facts he or she must judge the various special claims on their
merits in relation to the job in hand.

Skirting heaters emit a certain amount of radiant heat,
though the proportion is much lower than in the case of
radiators. Both because of this and because of the rising convection, it is not advisable to stand solid objects permanently
in front of the heaters at close range.

Convectors

Convectors form a class of appliance whose heat output is
almost entirely in the form of a warm air current. There are

two forms of convector, natural and fan assisted, and they deserve to be considered separately. The natural convector resembles the radiator in size and in being wall hung. It is particularly useful in situations where a high radiant factor is not required, or where a high convected output is best. A library or an art gallery is an example, requiring a good level of

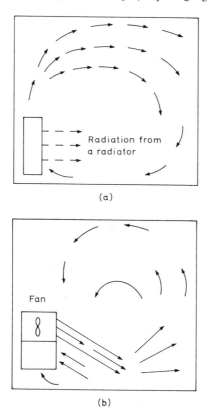

(a)

(b)

Fig. 2.19. Convected warm air pattern from a radiator or natural convector (a) and air travel pattern from a fan convector (b)

ambient warmth but absence of high spots caused by direct radiation.

The convector is capable of giving an output well above its rated value, since it can be connected to a pressurised system working at a temperature in excess of that which an atmospheric system can achieve. The front face remains at a safe temperature.

The detailed design of convectors varies considerably between makes, but all those worth considering have in common that they should be able to claim MARC approval.

The low level of radiant heat output of a convector brings another difference from radiators. We have been at pains to stress that radiators must not be crowded by furniture, which would intercept the radiant heat, with loss of useful efficiency and possible damage to the furniture. The only prohibited area around a convector is above it. For instance one would not hang a picture or a mirror over a convector (*Figure 2.19a*).

The fan convector is very different, because of its powered output (*Figure 2.19b*). It is usual to fit this item within about 0.3m (1ft) of the floor, allowing space for the circulating air to return to the lower half of the unit. Then the discharge, which in some cases has some directional adjustment, is angled downwards at the floor some 2m (6ft) distant. In this way it effectively prevents natural layering, the stratification which allows colder air to collect at floor level. The fan convector correctly installed can promote the lowest room temperature gradient of all heat emitters, no more than 2 or 3 degC difference between head and foot level.

To be able to do so it requires two conditions. First, it must have an uninterrupted distance in front of it of at least 3 m and preferably more, so that the 'throw' pattern can develop along with its secondary entrainment pattern. Then, it should be in a central position on a wall in order to obtain full advantage from the secondary entrainments which occur at both sides of the main air stream. A unit fitted near the end of a wall will lose that effect on one side.

The heat exchanger in a fan convector is a concentrated one. This means that a fan convector has an output equivalent to a natural radiator of at least six times its wall or picture frame

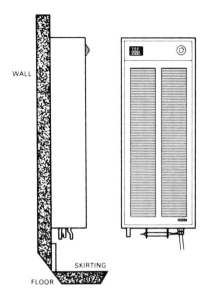

Fig. 2.20. A vertical fan convector (by
courtesy of Myson Group Marketing Ltd)

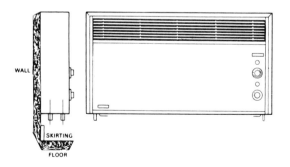

Fig. 2.21. A horizontal fan convector (by courtesy of
Myson Group Marketing Ltd)

area. It means too that the unit offers a quite high resistance to flow and must be fitted in a two-pipe circuit.

A fan convector is almost entirely dependent upon the fan for its output. Consequently its on/off control is very easily carried out, by means of an electric switch. Additional controls include a speed changing switch, to adjust the output to the *range* of temperature required, for example high speed for cold weather or quick warm-up, with medium or low speed for milder times or when the room is just about up to temperature. The final adjustment, to achieve a close control of room temperature, is brought about automatically by a built-in thermostat which senses the temperature of the air returning to the unit. Best running results are obtained by using the speed change switch so that the thermostat keeps the unit running most of the time. It is for example better for room temperature control to have the unit running 20 minutes out of every 30, than 10 in 30.

The aim in this chapter has been to present a broad picture of what goes to make up a wet system (excluding the boiler, which calls for special treatment). Certain patterns emerge. For instance:

A two-pipe system is better than a single-pipe system, but either is vastly better than a gravity system.

Except in a couple of small areas in Britain where the water is relatively harmless, no one should be allowed to fit a direct water system for heating.

Radiators, convectors and fan convectors can be installed relatively cheaply in many properties, using the close-coupled type of system.

The best temperature gradient results come from skirting heaters and fan convectors.

Fan convectors use very little wall space, are recommended for living rooms, not bedrooms (on account of the slight 'click' as the thermostat switches).

Radiators remain simple, easy to clean; their radiant output of some benefit.

There is room for all types of heat emitter, not only in the market but in any one house. For example, matching unit to duty, one could choose a fan convector for the living room, radiators for bedrooms, a natural convector for the hall.

3 The Boiler as Part of the Wet System

To repeat an earlier statement, the precise type of boiler or the fuel it uses is rarely a matter of any great importance, except to those who sell fuel. The only example of boilers which are not completely interchangeable is that of a system of heating only (with no domestic hot water) having all the heat emitters capable of shutting down at the same time. Such a system could not have a solid fuel boiler, which *must* have a buffer, usually the hot water cylinder, to mop up surplus heat during the slowing down period.

Pressure jet oil fired boilers are not often tolerated indoors, and to that extent are not fully interchangeable with other types. But that is only because of their operating noise, and not relevant to the present context. It is convenient when listing the principal types of boiler available to put them in fuel categories.

Gas: There is the free standing type, with smart casing if it goes in the kitchen, without case if it will go under the stairs or in the attic. The back boiler is allied to a gas fire and is not seen, but is generally much the same boiler as the free standing one, perhaps shortened. The wall hung boiler represents a new breed, with its own technology.

Oil is represented mainly by free standing types, of which there are three: pressure jet, wall flame and vaporiser, depending upon the type of combustion. Oil is taking from solid fuel some of the job of heating combination ranges of the Aga type, and there is even a unit fired by oil which closely resembles a gas fired back boiler with fire.

Solid fuel i.e. a grade of coal (see Fuels, Chapter 7), coke or prepared smokeless fuel. Without access to statistics, if such exist, it is probably true to say that there are more back boiler type appliances than any other type using solid fuel domestically. The choice among bigger and more instrumented appliances is now quite limited, and for an assessment at any given time one should consult the Solid Fuel Advisory Service, Hobart House, London SW1. Solid fuel has many admirable qualities, but it lacks the ability to be instantaneously controlled which its competitors have, and it is not therefore always best suited to work in systems which depend upon quick positive control. In the open fire/back boiler combination, where the only control is a manually operated damper, it does well.

The open fire/back boiler, even the closeable fire with back boiler, has been victimised over the years. Far too often it has, no doubt because of its relatively low cost, become associated with other forms of cost cutting which in these enlightened times should not be tolerated. The most usual is the direct system which we have seen to be responsible for bad internal troubles inside systems. By all means have a back boiler unit if you wish, but do not skimp the system on that account. Also, be aware of what you are buying in quantitative terms. Remember, *as much* heat as you need?

A typical boiler for a typical house would be rated at least 13kW or 45 000 Btu/h for full central heating. A typical open fire/back boiler is rated at about 7.5kW or 25 000 Btu/h. It is unlikely that this or any solid fuel appliance will maintain its rated output through all the changes of fuelling and clinkering, burn up and run down of fire bed intensity and so on. So this unit is necessarily associated with an installation with limited requirements, perhaps partial heating, or selective heating.

Gas Boilers

To look at boilers in the sort of detail which might count when deciding to buy, let us start with that which is numerically most in demand — the gas boiler. If we are to look for a reason for this success, perhaps the foremost is that a gas boiler is easy. It is so easy that it will go for a year entirely unattended. No fuel to order, store, carry; no lighting procedure; rarely any noise or temperamental behaviour. These are points which will appeal to those who are planning for retirement and an easy life.

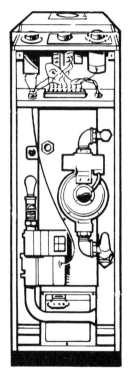

Fig. 3.1. A typical gas fired small bore unit for heating and hot water (by courtesy of Thorn Heating Ltd)

In spite of the growing variety, the standard gas boiler is still based upon a cast iron heat exchanger. This is the free standing boiler, with or without stove enamelled steel case, usually situated in the kitchen or under the stairs (*Figure 3.1*). There is no real limit to the places where a boiler may be placed, though bedrooms should be avoided on account of the disconcerting 'click' a thermostat can make in the quiet small hours.

It is customary to make the boiler casing to standard kitchen unit height. In spite of this being the subject of a British Standard, it is surprising how ideas vary about what an acceptable height is. The safe course, in order to match kitchen furniture already in place, is to determine the actual height. A short boiler can be stood on a metal plate, or tile, to increase its height.

Most manufacturers try to keep the back-to-front dimension down as well, still pursuing the elusive 'kitchen unit' standard. But the real stress is laid upon width, and with certain essential components to be housed it is obvious that the makers cannot go on cutting down all three dimensions. So, in the end, it may take some juggling to get the kitchen looking uniform.

A word of warning against too much emphasis upon neatness of packaging of kitchen cabinets etc. with the boiler. Read the boiler instructions carefully. It is very likely that there must be 50mm (2in) space at one or both sides. Some boilers take their essential air in at the back, and it is therefore vital for ample air to be available at the back of the boiler. The flue, if it goes away from the back, must not be in contact with any combustible material. The free space in front of the boiler, if necessary through an opening door, must be equal to the total length of the boiler, so that the burner may be withdrawn during servicing.

It is *not* a part of the responsibility of anyone employing an installer to decide between the merits of conventional and balanced flue. This is a matter which tends to decide itself, in relation to the site. If there is a good flue or chimney, and if the boiler may conveniently be joined to it and be acceptable in that position, then a conventionally flued model is the obvious choice. A balanced flue should not be regarded as an equal alternative, but as the second choice if a conventional flue is not available. This advice is strengthened by the fact that balanced flue models cost more to buy. Chapter 6 deals with flue characteristics but it is appropriate here to note that a balanced flue appliance takes its combustion air direct from outdoors, making no demands upon the room where it is situated. So if for any reason there is likely to be difficulty in getting an adequate air supply to the boiler, good conventional flue notwithstanding, then in that case a balanced flue model would be preferable.

The basic controls on a gas boiler are nowadays almost exclusively electrical. Despite the degree of dependence implied, the risk is small and the benefits, in crisp positive action, considerable. Thus, the boiler thermostat and the gas control valve, and the programmer if fitted, are all electrical. But it is comforting to note that the most basic safety device, the flame failure mechanism, does not depend upon current electricity. It works entirely off the pilot flame, which heats a small bimetal junction, generating an electromagnetic potential which holds open a key gas valve. If the pilot goes out that key gas valve cannot remain open, and nothing can open it. So however much the boiler instruments call for heat, no gas will flow unless the pilot is alight to ignite it.

A boiler which is engaged in heating is a part of a circuit which itself has to be controlled. There is a good deal of sense and usefulness in lumping most of those controls on to the boiler, so that it becomes the 'engine room' of the system. The commonest such addition is the programmer, whose principal duty is to control the starting and stopping times of the boiler, whether it is required to give domestic hot water only or a full service of hot water and central heating.

There are simple programmers with four or five programmes on offer, and there are magnificent units with a great many choices. We will not attempt to indicate which is preferable, and indeed it is a matter for personal choice, influenced perhaps by consideration of how the household operates in its coming and goings. The only pertinent observation to make from experience is that people with say ten programmes to choose from rarely use more than three or four.

A fully equipped boiler usually houses a junction box, into which are wired such external services as the pump and the room thermostat which, along with the clock in the programmer, controls its activity.

It is appropriate at this point to mention the use of the boiler thermostat. Some quaint ideas persist about this, for instance that using a very low temperature setting helps to keep the boiler from wearing out. Unfortunately for this theory the truth is quite opposite. A setting below 60°C (140°F) encourages the boiler to wear out since it permits condensation in the flueways. When direct systems were common, the problems connected with going to higher temperatures were just as serious. In hard water areas, the higher the temperature the more troublesome the scale deposition. That is why the 140°F was first selected, as a compromise. But now that systems are indirect the prohibition is removed, and we recommend the use of the thermostat for seasonal modulation of boiler output. In hard winter set it high, not less than 80°C (180°F), and indeed up to 90°C (190°F). This enables the boiler, and the heat emitters, radiators etc, to give their maximum output. In moderate winter weather this temperature may be relaxed, perhaps to the 70–75°C (160–170°F) range. The milder beginning and end of the heating season are the times for the lower setting. Such attentions are not strictly necessary, but one should never forget the usefulness of the upper range. Instead of getting annoyed with the inadequacy of the boiler, turn up the thermostat to make it work harder.

Ignition

A distinction must be made between first lighting, and the routine automatic lighting of the main burner which occurs under the influence of the boiler thermostat. The second is standardised, employing a small pilot flame which also heats the thermal junction − the flame safety device. The adjustment of the pilot is a part of the boiler design and should never be altered except by an expert. Spark ignition is becoming more common, but is not an alternative, since a given model or make of boiler has one or the other built in. Spark ignition saves the surprising annual cost of running a pilot. Pilot ignition has no moving parts to wear out. The lighting of the pilot introduces some variety of method, though too much should not be made of this, for it will normally occur only once or twice in a season. The room sealed appliance, i.e. balanced flue or Se-duct, is never nowadays lit manually. In anything like an on-terminal wind it is impossible. In the earliest days models were made for manual lighting, and users were advised to hang their hat over the terminal temporarily. But six floors up? With that exception, the main methods of lighting a pilot are:

(1) Manual, using a taper or match, never a bulky spill of paper.

(2) By glow coil, mains fed through a transformer. Many multifunctional controls operate at 12 volts, and the transformer handles everything. The glow coil method is losing popularity, partly because like all incandescent coils it has a measurable life.

(3) By piezo-electric igniter. The quite unexpected result of hitting a certain crystalline chemical a sharp blow is to produce a very high voltage, which can be made to jump a gap as a spark. This is the piezo effect and is built into a small triggering device in the form of a pull or push button. It is entirely independent of any outside source of energy. Piezo igniters are now being sold as hand held units, like the old flint lighter for cookers. They could therefore be used for manual lighting.

Servicing

The routine servicing of a gas boiler involves partial dismantling, removal of burner and of any cleaning cover necessary to enable the heat exchanger to be thoroughly brushed or scraped. A particular area for attention is in the gas injector and at the burner tips. The entrainment of air can lead to a collection of air borne dust and fibres, and a condition called 'linting' can occur. This interferes with the designed air/gas ratio, hence with good combustion. The prohibition against using a paper spill for lighting is to keep paper ash out of the burner.

All electrical connections should be examined, and so far as possible instruments and controls checked for operation. It is however usual in routine servicing to assume that if controls are known to be operating satisfactorily they are not in need of detailed attention. Servicing includes taking care of anything specifically mentioned as being in need of attention. Cover plates almost always have a gasket, of asbestos or similar, to form an air tight seal. These gaskets should be renewed.

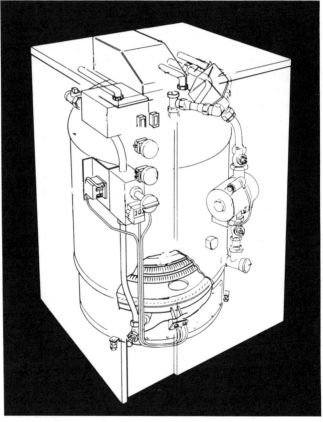

Fig. 3.2. The Worcester Delglo boiler, balanced flue model

A good service engineer will always leave the appliance at work, unless instructed otherwise, and will draw the owner's attention to anything he should know, such as a particular item which ought to receive attention at the next service visit.

Change and innovation are characteristics of a progressive technology, and there is always something happening in the total field of gas boilers and the like. Frequently it is not greatly significant, just enough to give a manufacturer a talking point, but one fairly recent change seems to be in a more significant category, in part because it has to do with that important household commodity, space. The innovators named it Mighty Mouse, though a change of company ownership resulted in the less memorable renaming to Worcester Delglo.

The ex-MM is based upon a gas fired boiler, output 13.4 kW or 45 000 Btu/h. Instead of serving a hot water cylinder in some remote airing cupboard, it incorporates its cylinder within the case, thus saving not only space but heat loss from pipework (*Figure 3.2*). More than that, it employs a larger than usual heat exchanger, giving a high heat make-up rate into a reduced volume of stored water. The makers claim that it will serve a bath every 12 minutes.

It has conventional and balanced flue models, and is British Gas approved.

The Gas Fired Back Boiler

Unlike the free standing boiler, which may be made in any size, the back boiler has limits imposed upon it by the aperture available. The free standing boiler is made right up to the generally accepted but arbitrary limit for domestic equipment of 44kW or 150 000 Btu/h. A back boiler is still struggling to beat 13kW or 45 000 Btu/h. But in terms of usefulness that is only a part of the story. The gas fire will almost always give ample warmth for the room in which it is situated, and the total effective output of the unit could be placed at say 4.5kW (15 000 Btu/h) higher.

The problem of how to get more than a pint into the pint pot of a British Standard chimney opening is almost the only technical problem left. Details of flue sharing with the fire, remote lighting of the pilot or burner, controls for the boiler incorporated in the fire, have all been satisfactorily settled. It must be noted however that the pairing of a fire and a boiler is fundamentally work for a manufacturer. If you have a gas fire you cannot go out and buy a boiler, and couple the two together. In case it is not obvious, this application is for conventional flue only, and it is noteworthy for two reasons. First, it makes the most of the limited space in modern dwellings, by taking no space at all. Second, it forces upon the user a source of radiant heat. This is in some respects superior to convected warmth though impractical as a source of 'central' heat.

All the comments already made about ignition and servicing apply, with (in the case of the latter) the extra proviso that the fire must be removed (and so serviced) in order to get at the boiler.

Wall Hung Boiler

A development even more recent than the back boiler is the wall hung boiler (*Figure 3.3*), another concession to the urge to save space. Weight was very much in mind at the start, and in some cases no doubt still is. Cast iron was abandoned in favour of alloy coated copper for the heat exchanger, and a fair comparison would be with an oversized Ascot type water heater. With this went very small waterways and an entirely new concept of urgency, necessary when applying a lot of heat to a little water. This had to be kept moving rapidly, and so was continuously pumped. A change in control methods had to be made, to ensure that the burner would not light unless the pump were working, and that the pump would stay working for long enough after the burner went out to purge the 'after heat'.

But now ways have been found to blend the old technology into the new package, and we have wall hung boilers which have cast iron heat exchangers and are suited to gravity circulation as well as to pumped primaries.

It must be kept in mind however that though the conditions may be relaxed they are most carefully worked out. When fitting one of these, or indeed any boiler, the maker's instructions should be read and followed very carefully. These

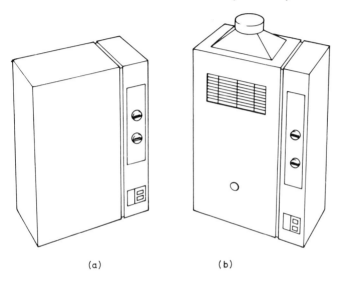

(a)   (b)

Fig. 3.3. Typical wall hung boilers, (a) with balanced flue, while (b) has a conventional flue

instructions are a part of the official approval and always a condition by which the guarantee is valid. In other words if you do not fit it as instructed you cannot expect any sympathy if it misbehaves.

Boilers on Bottled Gas

In theory – and we stress that point – there is no reason why almost any gas appliance ever devised should not be made in a bottled gas version. But in practice the list is quite short, for two reasons. One is that manufacturers are under no obligation to work out the necessary design characteristics in fulfilling their prime obligation, which is to the approvals authority, British Gas. Some do, and no doubt other makers would if they could command the very considerable work force needed. A second reason may be that conversion characteristics have been worked out as a laboratory exercise, but no parts manufactured. If anyone is interested in converting an existing appliance, or is particularly taken by one model of a new one, the only sure way to find out if parts or even information are available is to ask the technical staff of the manufacturer concerned. Given the information, the gas supplier might be able to do the conversion.

In spite of the difference in combustion and combustion equipment, bottled gas appliances are operationally similar to those using natural gas in all relevant respects. The technicalities are already built in and do not concern the user. Installation might be slightly different in that the required size of the gas pipe supplying the appliance can be smaller for bottled gas.

Some notes on bottled gas appear in Chapter 7. The only query likely to arise concerns the size of storage container, which may depend on the number and type of appliance to be supplied. This is best resolved by taking the advice of the supplier, which will be based upon national experience. The aim is to strike a balance between keeping the cost of hire down, and taking advantage of any bulk discount obtainable − and getting a reasonable time of running out of one container.

Oil Fired Boilers

Perhaps the most important distinction between types of oil boiler divides those which can be used indoors or out from those which are best suited to be away from the house. The latter are the pressure jets, and perhaps the name jet will suggest why this type of burner is best given a place of its own to operate in. Of the rest, those which depend entirely upon natural operation are silent, the rest make some noise however slight. It is always worth asking, in any given case, whether there is a history of the model making more noise as it ages and wears.

The principal types of vaporising burners, as they are called, are the wall flame (*Figure 3.4*) and the pot type, for natural or fan assisted draught. The wall flame and the fan assisted pot type must have electricity, but it is usual nowadays to electrify all types, because as in the case of gas it makes control so much easier and so much more standard.

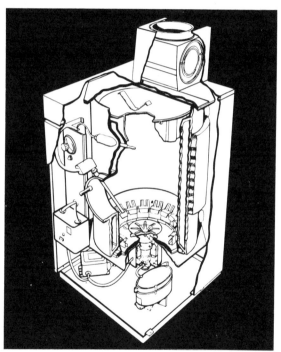

Fig. 3.4. A typical wall flame oil-fired boiler

Another common feature of these boilers is that they burn kerosene, which is 28 sec or Class C fuel oil. They operate in a routine way under the control of a boiler thermostat, and it depends upon the detail of the controller whether the burner works on/off or on high/low flame.

The incorporation of other controls, a clock or programmer for instance, present no difficulty so long as the boiler is electrified, for controls of this kind simply control electric circuits. A clock, for instance, is only a means of switching on or off, and so stopping or starting an operation.

We will not make a serious attempt to describe how these boilers operate. They are not suitable subjects for amateur adjustment or even servicing, and so the details are not relevant. It should be sufficient to know that all the vaporisers have a method of converting the liquid fuel to a vapour, in which form it burns. The method of conversion differs considerably from one type to the next. It is most important that the boiler is not neglected. For instance signs of sooting might indicate that a primary air port has become blocked, perhaps by something falling across it. Irregularities not understood should be reported, and as with gas boilers it is desirable to have an understanding, if not actually a service contract, with a trusted engineer.

Not all models of this type of boiler are available over the entire domestic range, nominally up to 44 kW or 150 000 Btu/h. The starting range is about 8 kW or 27 000 Btu/h. Forced draught vaporisers go up to about 23 kW or 80 000 Btu/h, and there are wall flame or rotary vaporisers up to the maximum.

A point to watch for and to get settled at the earliest stage concerns the type of combustion control and the maker's instructions about the system. Boilers of the vaporising type which have on/off control are in general in line with any other instantaneously controlled boiler unit, such as gas. They are safe for use in a 'heating only' system if necessary. But boilers which have high/low flame control, in which the low flame may be compared to an exaggerated pilot, generate an amount of heat in their 'at rest' position, which must be dissipated if overheating is to be avoided. It is customary to employ this surplus warmth usefully, through the primary circuit which feeds the domestic hot water cylinder, the cylinder being an essential part of such a circuit. The heat emitter is a towel rail or small radiator, usually in the bathroom, which is always in circuit. Being in the hot water primary it is independent of any circulating pump which may be on the heating circuit proper. The need to know about any such requirement *before* ordering materials and running pipes, will be obvious.

We have already described the pressure jet boiler as a type unsuited to being indoors, and this is bound to bring pained reaction from some manufacturers. Operational noise occupies a major part of their development engineers' attention, and someone is always claiming to have beaten the bogey. But in the end it is the owner's sensitivity of hearing which settles the matter, and we would advise anyone in doubt to ask for a practical demonstration. But let us add one other note of caution. The site can sometimes influence the nature and level of noise. There are cases of a pressure jet being connected to a

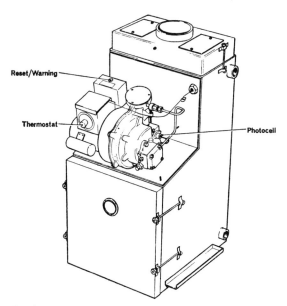

Fig. 3.5. A down fired pressure jet oil-fired boiler, without case (by courtesy of Thorn Heating Ltd)

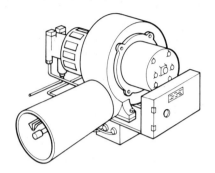

Fig. 3.6. Typical pressure jet burner

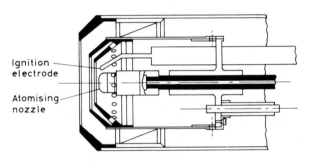

Ignition
electrode

Atomising
nozzle

Fig. 3.7. A typical combustion head used in a pressure jet burner

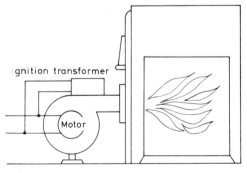

Ignition transformer

Motor

Fig. 3.8. How a pressure jet burner fires

chimney which runs by a bedroom, where the flue gases have to make a sharp change of direction. The effect, particularly in the quiet of night, is of distant battle, muffled but insistent. It is a clear warning to pay attention to the flue as part of the whole installation.

Pressure jets differ from other forms of combustion, in that the flame is not divided between a number of burner jets or spread over an area. It is concentrated in one jet, rather like the water from a fireman's hose, and like that example it issues under pressure. The fuel oil is ejected under pressure from the pump which is incorporated in the burner, and at the point of release is atomised by being given a swirling motion as it passes through a very finely machined and calibrated nozzle. The combustion air, also under pressure and controlled, is admitted and admixed in a very precise way, within the burner assembly. By choice of nozzle, by adjustment of the pump pressure and the air inlet damper, a high degree of precision is obtainable. One of the principal objects of regular service is to ensure that the set conditions are maintained: to keep the nozzle free from foreign matter which might have got past the filters; to see that there is no interference with the air ways.

Pressure jets are on/off burners, and ignition is by spark from a pair of electrodes bridging the exit of the fuel nozzle.

The brain and heart of a pressure jet is the control box. It houses the boiler thermostat, for instance, and when this is satisfied it cuts the current to the burner pump and fan. When the thermostat again calls for heat it sets the spark ignition going, as a preliminary to starting oil pump and fan. If the burner does not light it persists for a given number of seconds, then stops. There is then a pause, known as a purging period, to allow unburnt oil vapour to escape. The control box will then initiate another firing period of similar length to the first. If this fails the burner will shut down, and a red light will show. It is a signal to see what might be wrong, and nothing can be restarted until the user has manually reset the burner by pressing a green button.

At this stage someone is sure to ask how it is that the control box knows that the burner has not lit. Well, it has an eye, a photo-electric cell which is inserted into the combustion chamber. When it sees a flame it generates a very small electric current, enough to inform the control box. If that information is missing the control box reacts in the way described. If it does, the first and most obvious step to take is to make sure that the 'eye' is clean. Quite often a shut-down is due to nothing worse than a thin film of carbon or soot on the glass. But, such is the overall reliability of the device, the second most common cause of flame failure is running out of fuel. This might be due to a blocked filter or pipe line rather than an empty tank.

Pressure jet boilers are generally less sensitive about flue conditions than the vaporising types, for the very good reason that they create a pressure in the combustion chamber to start the evacuation process going.

The smallest pressure jet boilers are in the 14 kW or 50 000 Btu/h range, but they then go on into sizes far above the domestic range. They belong therefore to the larger installation.

Pressure jet burners may be specified for one class of fuel only, i.e. Class C or D, which is kerosene or gas oil. Or they may be classed as suitable for either. In that case it is necessary to see that all the conditions relative to the class of fuel being used are observed. These conditions usually refer to the rating of the nozzle, the oil pump pressure, and then during setting up the correct adjustment of the air damper to go with the oil input. It must not be expected that a change can be made from one grade of oil to another without such adjustments.

If the fuel oil store happens to be buried, or at a lower level than the burner, the pressure jet burner offers one advantage. When considering fuels reference is made to the two-pipe system for getting a supply of oil to the burner. The pump for the purpose is the oil pump on the pressure jet burner, with only a simple adjustment to be made during installation.

Combined units

Oil, like gas and solid fuel, has its hearth fitted unit with a fire in the front and a boiler at the back. The similarity with gas is the more obvious, since the fire resembles a gas fire in appearance. A typical model is as described below.

The boiler is a natural draught vaporising type, with high/low control, and the notes about installation and dissipation of surplus heat apply to it. The total rated output of the model shown is 10.9 kW or 37 400 Btu/h, of which the amount going to water may be varied, from 5.1 kW to 9.3 kW, the rest of course going to the room as radiation and convection. The method by which this ratio is determined could not be simpler.

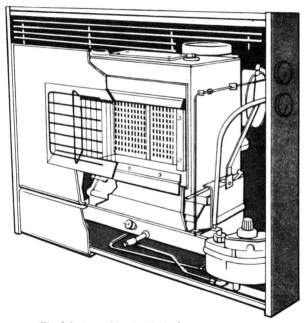

Fig. 3.9. A combination boiler/room heater unit

By means of a knob at the side, the angle at which the radiant section rests is varied, to bring more or less of it into direct heat. The whole of the section is glass fronted.

Because of the limitations imposed by the chimney size, this type of unit is unlikely to exceed 12 kW or 40 000 Btu/h rated output. Within that total will be ample output for most rooms in which the unit is likely to be situated. The oil supply is usually piped in from outdoor storage. But it might interest in particular anyone seeking a 'can take it with you' unit to find one which has an incorporated fuel tank.

Flue Stabiliser

A flue stabiliser is usually in the form of a flap damper which is free to swing, from shut to some way open, under the influence of changing chimney draught or 'pull'. Its purpose is to stabilise the amount of draught at the base of the chimney, when the natural draught emanating from the top of the chimney is fluctuating. A surge of draught causes the damper to swing open, and air enters through the open damper, sufficient to satisfy the extra demand caused by the draught increase. In consequence the extra demand is not passed on to the base, and so to the boiler or other appliance fitted to the chimney.

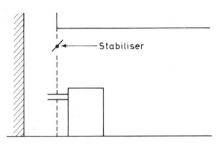

Fig. 3.10. Draught stabiliser fitted in boiler room

Oil fired appliances in particular work best in stable draught conditions, and of those the vaporising type are most sensitive. They are generally fitted with an integral draught stabiliser in the flue offtake (*Figure 3.10*), and this should be able to swing freely.

Pressure jet boilers, due to the start they get from having a slightly pressurised combustion chamber, do not often have a built-in stabiliser, and makers commonly state that in normal conditions none is needed. But in extreme conditions, such as one might meet on an open site in, say North Cornwall, high winds might cause high and fluctuating draughts in a chimney. In such cases, and in cases where vaporising boilers are not already equipped, a separate stabiliser would have to be fitted in the chimney.

When this is done, *it is imperative* that it shall be fitted in the same room as the boiler or appliance. It cannot be for instance in the room above. It is usually put up near the ceiling. The reason for this condition is that it must share the conditions which apply to the burner, on both sides of the damper, inside and outside (*Figure 3.10*).

Solid Fuel Boilers

The breadth of choice of heating equipment with solid fuel is more than with other fuels. In consequence of this there is a greater responsibility upon the seller of apparatus to declare the maximum performance, or rating, of every appliance. It is when this responsibility is avoided, and too much emphasis placed upon the simple evocative phrase 'central heating' that troubles can start. The law recognises that the buyer has a duty to look after himself, and in this market a potential buyer should always enquire, first, what is the official rating of an appliance.

Then, except in the case of automatically fed hopper charged boilers, he should allow for the difference between test conditions, charging every 1½ hours, and the real life conditions into which the appliance would fit. It would not be unreasonable to take 10% off the rated output for that.

Because the result of such a preliminary enquiry will be to narrow the field of choice at the outset, let us start there.

(1) The open fire with back boiler will have a boiler output in the region 3 kW or 10 000 Btu/h. If described as high

output this could go as high as 7 kW (23 000 Btu/h) or even higher. This is 'rated output' of course. Except in the smallest house this is for partial or background or selective heating only.

(2) The closed or closeable room heater with back boiler. We must consider this working as a closed unit since the opening of the front causes a significant drop in performance. The maximum rating for such a unit is 13 kW or 43 500 Btu/h which if sustained could satisfy a good proportion of dwellings. To that output must be added the convection and radiation from the heater which is probably adequate for the room where it is situated. We would recommend that for practical purposes the output be taken to be below the rated figure, to avoid disappointment.

(3) The free standing manually fed boiler. This is probably the type with which older people are most familiar, perhaps when it was in the form of a cylinder, of cast iron with a lid on the top. Originally coke boilers, they will burn any smokeless fuel, the correct size grade being that recommended by the manufacturer. The use of bituminous or house coal is allowed only if the firebed is specially formed as a 'Smoke Eater'. This type of boiler is frequently used as an incinerator of kitchen rubbish. We cannot stop that practice merely by disapproving of it, but may we urge moderation. Years ago we were puzzled by the lady who complained that whenever she burned cabbage stalks she got excessive clinker formation. It did not seem like a common property of cabbage. After a time it was traced to the fact that the cabbage stalks tended to damp the fire down, and the owner opened the damper wide to brighten it up again. Hence the heavy clinker. Manually fed boilers for domestic use fall into a category roughly between 5 and 15 kW, or 17 and 50 000 Btu/h rated output.

(4) The free standing hopper fed boiler is way ahead of the others in operational ease and efficiency (*Figure 3.11*). It provides the working conditions in which the purest of the solid fuels, anthracite, will burn, and in return gives steady output and very limited interference from ash. It is a demanding

boiler, in that it must have a good flue and that size of fuel is critical. The price of anthracite is matched to the size, and most of the hopper fed boilers currently available use peas or grains. The density of packing of these sizes means that a small electric fan is needed to force air through

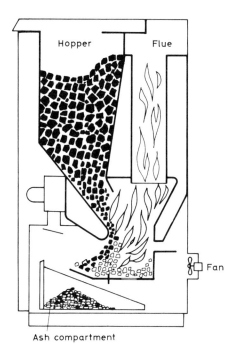

Fig. 3.11. How a hopper fed boiler works

the fuel bed. This offsets any saving on fuel cost but gives an extra measure of control over combustion. One side benefit is to extend greatly the length of time the boiler can stand idling, without the fire going out. A clock is wired into the fan circuit, timed to cut in for a short period at intervals, in order to brighten the fire. If there is a natural draught boiler available at the time of going to print, it will use a larger fuel, probably beans. Any boiler in this class which is allowed to burn bituminous fuel will have a Smoke Eater grade.

The frequency of filling the hopper, and of removing the ash, are competitive features between makes. But bearing in mind that there is no poking or riddling to do — and indeed it would be wrong to do it — then attention once daily should not be excessive. In most models it is arranged that the ash when withdrawn is nearly cold, and so not giving rise to air borne dust.

These are the larger boilers, from about 13 kW or 40 000 Btu/h upward and into the commercial sizes. They are therefore readily available for full central heating in any domestic premises.

We have already commented upon what might be called the social grading of these types of boiler, but it bears repeating. There is *no* justification for the belief that the first two, and perhaps the third, can satisfactorily be connected to direct water systems. The need for an indirect system is as urgent for them as it is for the last, the hopper fed boiler. It is the heat, not the type of boiler, which damages the system.

Despite their wide differences, the four types of boiler have two things in common:

(1) They all have to be lit manually. This may not be the paper and sticks method, and it is quite common to use a gas poker, or an electric poker which generates a stream of very hot air to blow on to the fuel. Or again there are patent fire-lighters. The manufacturer will almost always have something useful to say on the subject, in his instructions.

(2) They all have the phenomenon of response lag, which causes unwanted heat to be generated after the command to stop. Consequently they all must be provided with a buffer or heat leak. In most cases, certainly where the outputs are in the lower range, the hot water cylinder will absorb the normal surplus. But in case of doubt, and for

larger boilers, it is desirable to include a radiator or towel rail in the hot water primary circuit, to be left on continuously.

At the top end of the domestic range is a class of boiler which is industrial in design, and unquestionably needs to be given a boiler house. It requires an independent hopper, facing the boiler, the two being connected by a screw feed mechanism which provides a slow but constant supply of fuel to the base of the fire. This is called under-feed stoking.

The smallest size likely in this would be about 24 kW or 80 000 Btu/h. Apart from the amount of space needed, it may be claimed for this type of boiler that it does away with even daily attention such as a gravity fed hopper boiler needs. Offsetting that is the cost of operating the screw feed.

In comparison with other fuels, the lighting of solid fuel is a chore, and one to be avoided for as long as possible. There is the added disadvantage that an appliance which has gone cold takes a relatively long time to warm up again, because the fire is slow in getting hold. It is of considerable importance therefore to know how long any appliance will keep going between refuellings, so that it may be left and not go out. The obvious and recurring period of neglect is overnight.

Though it is clear that the period will depend upon the rate of burning, whether the boiler is working hard or idling during the period, it is usual to assume that we are considering an idling period, overnight being again typical.

The hopper fed boilers present no problems in fuel availability, since the hopper capacity is almost always capable of riding over at least a day, usually longer at low or nil load. In fact the problem with such boilers is more to prevent them dying of inaction.

The manually fed free standing boiler and the closed room heater are usually designed to have enough room for fuel so that with a nearly closed damper they will idle on for many hours.

The open fire rarely has any such provision, even though the need for continuity of combustion is there. Consequently the makers, if in doubt, supply a deepening bar, a removable plate which is slipped in at night to increase the height of the front of the fire, so creating a larger pocket for fuel. The nature and usefulness of the deepening bar if supplied are examined by the approval authority, and if satisfactory are included in the approval. Intending purchasers might like to note that they have a perfect right to see the precise terms of any approval granted to an appliance and the seller should make this available to them.

It is usual to remove the deepening bar during normal usage hours, not least because it will shield off some of the radiation from the room. Along with its use as a means of prolonging an active fire goes partial closing of the damper which is below the fire bed. This greatly reduces the amount of air able to get under the fire bed, while one role of the deepening bar is to deflect off the air which otherwise would sweep across the top of the fire — both being responsible for some fuel comsumption.

Many modern units have one more damper, fitted in the throat or go-away flue of the appliance. The purpose of this damper is to control by throttling the amount of draught or 'pull' which the chimney is exerting upon the fire, which in practical terms means the amount of air it is persuading the fire to use. One operation in preparation for idling is to part close this damper, so reducing the demand made by the chimney. All the preparations are designed to reduce activity, and not for nothing is a fire in this condition said to be slumbering. The maker's instructions will generally give an indication of the best settings for dampers, though it will be obvious that no absolute values can be given since chimney draughts vary from place to place and from time to time. Fully shutting dampers will almost always result in the fire going out.

Electric Boilers

Electricity does not come readily to mind when talking about boilers. Nevertheless there are two kinds of electric boiler which must be mentioned in the domestic field.

One, the electrode boiler, seems never to have become a serious item on the market, and may be disregarded. The other, known as the Centralec system, is available and would be of particular interest to those who are debarred from having any kind of flue on their premises. Centralec is basically a storage heater of the storage radiator type, which instead of passing its stored heat into the air passes it into water, under control. The water is, of course, the wet heating system, which after it leaves the Centralec unit is exactly the same as any other wet system.

The heater being of the storage pattern qualifies for the use of cheap off-peak current, which is charged at off-peak hours. The capacity of the unit is then made to match the heat requirements over the whole 24 hours. The unit may be obtained with a fitted circulating pump, or an external pump may be used.

One of the advantages of a wet system over a storage radiator system is in greater precision of control over heat output. In Centralec we find electric storage allied to finer control, another advantage.

The Size of the Boiler

A custom has grown up, from father to son as it were, of calculating the size of boiler required for a job, and then adding 10% or even 20%, and we have known even more. It is a practice so well dug in that even text books quote it. But it is neither necessary nor good practice. It arose originally because the waywardness of the solid fuel boiler in the user's hands was recognised, and an allowance made for that. As a cover against bad calculations it was welcomed by some workers, who usually excused it as a hedge against extra bad weather.

We have already mentioned that in the case of manually fed solid fuel boilers one should take a conservative view of the real output in comparison with the rated output. But that is the only case. In all other circumstances oversizing – for that is what it is – is wrong, leading to waste of money and early wear on the boiler. Since the greater part of the heating season is far from being as severe as that on which the calculations are based, we already have a situation, with a boiler of correct size, where it takes a number of rest periods during working hours because it has done its job quickly. Every time it goes off, which it might do for a total of say 30 minutes every hour, things start to cool, and heat is dissipated mostly uselessly. When it goes on again that heat has to be made good, and all this counts against the running cost. Consequently, to make the boiler even bigger is simply to add to the down times, the cooling and the loss.

To anyone tempted still to think of the extreme case of bad weather let us say this. Life is almost wholly made up of quite ordinary weather, bad though it may be. When the wolves are howling outside it is a crisis situation in which you can afford to make a temporary concession. Decide to see it out by having heating in only say two rooms, instead of four or five or six. For that limited duty the boiler will be adequate. But do not live the rest of your life paying extravagantly for the remote possibility that some such crisis might occur.

4 Warm Air Systems

In the Introduction we mentioned the benefits of using air as the heating medium. There is some similarity between wet and dry systems, in that both are capable of being left to Nature to operate, or may be given a mechanical boost. The one fact to be remembered about Nature, apart from the fact that it costs nothing, is that while it is inexorable it is slow and comparatively weak. For instance nothing will stop warm air at ground floor level from rising to the upper floor. But shutting a bedroom door will prove a serious barrier, and it could take a week for the warm air to get through. That is hardly a practical proposition, but it serves to show incidentally that free flowing warm air is most likely to suceed with open planning.

In a dwelling so constructed that free flowing warm air can circulate, there could be a lot of sense, and of economy in first and running costs, in having such an arrangement – it can scarcely be called a system. For best operation it calls for a good standard of loft insulation, and for double glazing in bedrooms. Anyone who does not fancy a well warmed bedroom has only to keep the bedroom door shut to stay out of the circulating zone. We have seen figures of quite unbelievable accuracy for temperatures in rooms warmed by drift in this way, but they are all rather beside the point. The human body is not seeking a specific temperature, but a comfortable zone of temperature above the minimum level of acceptance.

The principle upon which all air heaters are based is that there is a heat exchanger. Fuel is burned on one side of it, and air passes over the other side, becoming warmed (Electricity is of course the odd man out, having no need to be segregated from the air by an independent heat exchanger.) The manner in which the principle is applied illustrates mainly the ingenuity of manufacturers. The commonest of the free flow units is still the so-called 'brick central', because that is what it is, both brick and central. It can house quite a large unit, which calls to mind that in the USA the air heater is called the furnace. No doubt the use of a brick chamber was first employed to make safety obvious, giving some thermal storage as a bonus. In later times units were built in which the chamber was more lightly constructed, for example of asbestos cement sheets. The cladding, brick or otherwise, is *not* a part of the heater. There is a flue, and grilles at high level to allow warmed air to escape. At low level, apart from the combustion air inlet, there is at least one inlet for cool air to be warmed, and a measure of control is exercised over the output by the extent of opening of lower air dampers.

In cases where the brick central was not situated in a way to give warm air an easy passage upstairs it became customary to add a short duct from the top of the unit to a convenient discharge point. The decline in popularity of the brick central is no doubt due to its massiveness, both real and apparent. But that apart it is capable of doing a good job.

The same function is being performed, for much less space taken, by an appliance which looks very like a large gas fire. Using fuel oil, which may be piped in or contained in an attached storage vessel, this unit must be connected to a chimney, so that choice of location is limited. Some models have a radiant section (like a gas fire), even a back boiler (also like some gas fires) but the bulk of their output is in warm air, discharged into the room where the chimney is. In some cases this warmed air is able to travel freely. Where it is not, its passage is aided by a simple extractor fan, let into the wall of the room at high level, and discharging for preference into a hall or passage way by which the warm air may go on its travels to the rest of the dwelling. Reports from users seem to be predominantly favourable.

For warm air heating which qualifies as a system we must turn to ducted air. It obeys all the rules laid down for heating

description, in that it may be a full heating system, or partial, or selective or background. The full range is available only to full duct systems. A stub duct or abbreviated duct system cannot fulfil the requirements of full central heating in terms of individual room control.

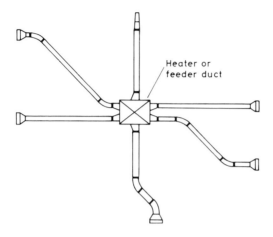

Heater or feeder duct

Fig. 4.1. The radial duct system, for cases where the heater can be centrally located. No branch pipe should exceed 6 m (20ft) or leave more than two bends in the run

The heart of a ducted system is the fan, almost always incorporated in the unit. The fan may be compared directly with the pump in small bore systems. It provides positive flow, speed, assurance, and it frees the system from the annoying limitations which Nature imposes upon a system without artificial power. Notably of course, in natural systems warm air must always be travelling in an upward direction. With fan power, it will travel for long distances horizontally, and even go downward, as into an underfloor duct system.

Ducting is large, not often an item which can be added to an existing house without becoming a nuisance. Consequently a full duct system is usually built into a new house, and the architect or builder should be given early notice so that he may plan it in. Ducting may be run under floors or in lofts, in walls. It may house the required cross sectional area in various shapes. A 6 x 8, for instance, giving an area of 48, may be made as 12 x 4 if it has to disappear in a four inch wall. Although the object is to make duct runs direct, short and economical, all flow ducts are insulated to guard against much unplanned heat loss.

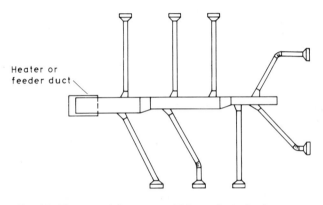

Heater or
feeder duct

Fig. 4.2. The stepped duct system. This may be in the elementary form shown, with only one main duct. Or the heater may supply more than one main duct and, up to the capacity of the heater, form a radial system as above. The difference between that and the radial system of Fig. 4.1 is in scale. Note that the offtakes from the main duct are 'swept' to assist flow in the changed direction

Another feature of long flow ducts is that the cross section undergoes gradual reduction. Each length of duct is sized according to the amount of air it has to carry, in order to maintain a steady air velocity. This is necessary in order to achieve a certain velocity of air delivery at the outlet points. It will be seen that as draw-offs are made from the main air duct, the quantity diminishes and the area of the duct must be reduced to suit the new reduced volume. Such calculations are made on the basis of all discharge points at work. In any other event if the discharge rate seems too strong it may be adjusted at the damper which controls the discharge.

It has already been suggested that an arrangement by which warm air goes where it pleases does not constitute a system. A system is a logical development, with beginning and end, and the logical development of a ducted system is to return the air, or at least a good deal of it, to the starting point. Thus, a ducted system includes a return air duct, and it is upon this recycling that the economy of the system mainly depends. By comparison with the flow ducting, a return air duct is a simple affair. Usually it collects from main points only, is not insulated, and not made leakproof to more than an elementary degree. The exception to the last point occurs within the heating chamber, or where the heater is situated. There the return air duct must be very effectively sealed in itself and into the heater. If the heater is a conventional flue type it relies upon access to a continuous supply of fresh air for combustion. It cannot tolerate strong competition such as would occur if there were an aperture in the return air duct, already under strong fan suction.

More than that, if incomplete combustion were to occur, with probably the formation of carbon monoxide, it would get into the duct system and be distributed throughout the house.

The fact that a duct system necessarily forms a communication channel from room to room has not escaped the attention of fire prevention officers, and it is advisable to check with the local authority in case their interpretation of the Building Regulations bears upon your proposals.

Sound also passes readily along this type of communication channel and this seems to be particularly true of the return air duct, no doubt because it is intentionally straight. Fan noise will travel in it unless checked, and the most effective check is a bend.

At this point we should mention that anyone seriously intending to install ducted warm air would need to go into the subject in greater depth than this book can, for ducted warm air has never been regarded as a very suitable matter for the amateur. Perhaps, with its new building context, it rarely

offers the chance. However, all the leading manufacturers of
warm air units, who together form the Warm Air Group of the
Society of British Gas Industries, have collaborated in producing
a design manual (published by Ernest Benn, 1976), which sets
out the current British practice.

Anyone who objects that it is a gas document should be
reminded that systems are more important than fuels. The
majority of the manual concerns duct systems, which do not
alter just because the fuel does.

We have seen how main ducts are stepped, i.e. progressively
reduced in cross sectional area as the volume diminishes, in
order to maintain a reasonably steady air velocity. If more than
one main duct is run from the plenum (which is in effect the
distributing box attached to the heater) then the design intention should be to give roughly equal duty to each main duct.

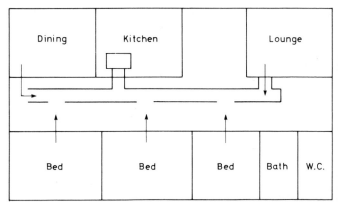

*Fig. 4. 3. Typical return air duct in a three bedroom bungalow. The
duct runs wholly in the passage. Grilles at high level allow air to pass
from bedrooms and dining room to passage, and grilles let into the duct
continue the communication. The lounge, having more to contribute,
and generally for longer, is given a stub duct connection. The unit is in
the kitchen in this case.*

*Note that under no cirumstances is provision made for pick-up from
w.c., bathroom or kitchen. It will be appreciated that what enters the
return air duct will emerge eventually at registers, and the system is
not intended to distribute cooking smells throughout the dwelling*

This intention can rarely be achieved in practice, and it is customary to add the final adjustment, known as balancing the system, by means of dampers. The comparison with wet systems, and balancing valves on radiators, will be apparent. Sometimes the need for an incorporated damper will be obvious. For instance if a radial system were to consist of one short duct and a number of longer ducts of about equal resistance, the short duct would have an advantage, and this would require a damper to neutralise it.

On a more general level, it is good practice to include a damper in every register. (A register is an outlet with some form of adjustment; return air inlets and the like, plain apertures with no adjustment, are grilles.) Formerly this called for a separate device, a stack damper, but nowadays it is much more common to include a balancing damper in the total structure of the register. This is independent of any modulating mechanism, including an 'off' position, which is provided for the user to control. Balancing dampers once set should never need alteration.

Where to put Registers and Grilles

Registers or diffusers are the terminal fittings through which warm air is discharged into the room. They should bring about as uniform a room air temperature as is possible, by low speed draught-free air movement. Their relationship to the room, and if more than one to each other, is therefore of first importance. Their ability to project air sideways and forward is limited by the air velocity, which must not create a draught, but their influence is greatly extended by secondary means, by disturbance and by entrainment of air surrounding the ejected warm air stream.

As a convenient rough rule we may assume that one register will deal effectively with a room, or part room, which has a small to medium square plan area. This would include such rooms as 4 x 4m (12 x 12ft), 5 x 5m (15 x 15ft) or 4 x 5m (12 x 15ft). Rooms which are distinctly elongated (e.g. 4 x 7m,

12 x 20ft) or irregular (e.g. L shaped) are best treated by being visualised as squares, and each square treated separately (see *Figure 4.4*).

The decision whether to fit the diffusers in the floor, at low or high level in the wall, or in the ceiling, is in part dictated by local circumstances, in part by local custom. British preference

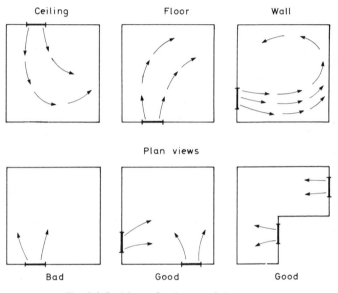

Fig. 4.4. Positions of registers and air patterns

has always been for the low level wall diffuser, and if this is arranged to give an angular downward deflection to the issuing air it can achieve very good mixing, with even temperature throughout the vertical gradient. It requires the ability to conceal the ducting in the wall, but that is a common condition. Floor diffusers require that an insulated duct can be accommodated in the floor.

High level diffusers, and in particular the ceiling type, are not for general application, since their natural trend would be to promote a great temperature gradient, hot at ceiling level and cold at the feet. They would be well suited to a place where the floor is already well warmed, possibly by being over another warm room.

It must be pointed out that the air flow from a register must be unimpeded, furniture being the most likely impediment. But in nearly all cases it should be possible to accommodate the furniture to the heating, not the other way round.

Return Air

The number of grilles used will be less than the number of registers, and the system does not even call for a grille in every room. But if a room is equipped with a register but no grille, then there must be a permanent opening at high level through which air can escape to the nearest grille. The typical site is one in which a central grille in a hall or passage collects from one or more adjacent rooms. The permanent opening is most easily provided by relieving the top of the door, though an opening may be made in the wall. No such arrangement must be made with the bathroom, kitchen and toilet, and if these abut a hall equipped with a return air grille steps should be taken to see that they cannot contribute accidentally to the return air supply. Their ventilation should be arranged separately, possibly via a window ventilator.

When grilles are fitted in the same room as registers it will be seen that the zone of negative pressure (the suck) in their vicinity will influence the air flow pattern. This fact can be used with advantage if the grille is situated with this in mind to create air flow in the required direction.

Since the object is to involve as much of the room as possible in the warming current of air, it may be taken that broadly the air will enter at one end of the room and leave at the other.

The Compromise System

We have considered the free flow unit, which uses no ducting; and the full duct system, which is best suited to new building.

It is often possible to obtain many of the advantages of the second, in application to existing property and with a minimum of disturbance. The compromise is called a stub duct system, and as the name indicates it uses only short duct lengths. These should be long enough to send the warmed air in the main directions it is required to go. The feasibility is very largely dependent upon the layout of the house, and the ingenuity of the designer. If there is more than one position where the heater could go, the object is to choose the one which gives most chances to run effective stub ducts.

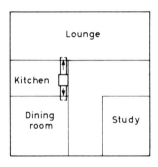
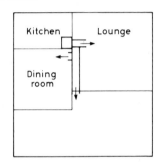

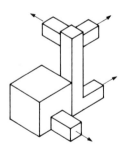
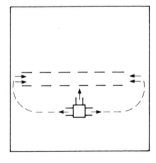

Fig. 4.5. Typical stub duct layouts for different house plans. The diagram bottom left shows a stub duct system on two floors, while the bottom right diagram illustrates the idea of influencing air travel from a stub duct system by means of the return air duct

We have tried to show, in sketch form, ways of achieving this in differing house plans. For houses of two storeys the object may be to transfer part of the air ouput to the upper floor by a single duct, then taking stub ducts from this riser to convenient nearby points.

The conditions imposed upon a return air duct are, as we have seen, not too onerous. A return air duct could for instance be run across a loft, which might well be unsuitable for the flow ducting. In such a case the return air duct could well be more extensive than the stub duct system, and so arranged that by putting grilles at negative pressure at the extreme ends to which it is hoped air will travel, it will induce air travel. *Figure 4.5* illustrates the principle.

Another very important matter concerned with ducting is fan power. Clearly it takes less fan power to operate a stub duct than a full duct system. Even nowadays, when the matter has been fairly well regularised, it is still desirable to ask, when buying an air heater, not only the rated heating power but also the fan power, whether for stub or full duct system.

Control of Warm Air Systems

In free flow units there is only the heater to control. This may be given automatic control in one or more respects, in addition to any fundamental controls, e.g. for flame safety, which may be incorporated in the burner assembly. A room thermostat, for example, may be situated in perhaps the living room, wired back to stop and start the burner. A clock will also decide when the unit is to be started or stopped.

Both these controls are applicable to ducted unit systems too, the room thermostat being more likely to give a sensible response in a system which is positively fed with predetermined amounts of warmth via ducts, than in conditions which can be affected by a draught from an open door.

It is possible to obtain thermostatically operated registers, though it is doubtful whether these contribute enough to the end result to justify their inclusion. The most important part of a duct system for the user, if only in the all-important matter of relating comfort to economy, is the register. It should become as much a routine to shut a register when leaving a room as it is to turn off the light.

Refreshing Warm Air

A warm air system was once regarded as one of diminishing vitality, in which the oxygen content grew less, the carbon dioxide, tobacco smoke and so on increased, presumably to a point at which it would not support life. To offset this, arrangements were made to introduce a controlled quantity of fresh air from outdoors into the circulation, and this would displace an equal volume of stale air.

Experience quite quickly showed that what might be called a laboratory view of the system at work had little relevance to what happened in practice. This is, to put it crudely, that a typical house leaks like a basket at most of its joints, floor boards and so on. The real problem became not one of introducing fresh air but of excluding too much of it, for it will be seen that fresh air is cold air and the need to treat it lowers the efficiency and therefore the economy of the system at work.

Although the introduction of fresh air receives a mention in the latest official document on ducted air design, our recommendation is to ignore it, unless you are convinced that you have a remarkably well constructed and weatherproofed house entirely free from draught leakage at any point.

The Nature of Warm Air Heating

Wet systems, we have seen, work ultimately by warming air, but it is an indirect process and, particularly in the case of radiators, the total effort is divided to give a radiant factor. By their nature they are, too, rather slow, even in spite of pumped circulation. It can take half an hour or more from starting the system before an improvement is noticed in the room temperature. Even then the convected or warm air part of the improvement lags behind the radiant one.

A warm air system is quite different. It is direct, the warm air coming 'from factory to user' with no intermediate process. It is simple in action, being convected only, and it is rapid. The room near to the register will begin to feel warm minutes after the apparatus goes to work, and it is quite possible to have the room up to temperature in about 15 minutes. But that is conditional upon limiting heat loss.

So in return for good insulation you will save money and will get the greatest benefit from a warm air system, in speed of response. The good sense and economy of insulation apply of course with equal force to wet systems. But for anyone specially interested in taking advantage of the speedy action of warm air it might pay to think about the detail of insulation. We shall go into the subject more fully later on, but a brief mention here is not out of place.

Cavity wall insulation is a very important part of house treatment, since it achieves a spectacular heat saving over the greatest surface area the house offers to the outside world. It is also very convenient. The cavity is a ready made former for the insulation, giving it standard thickness but not allowing it to take up any useful room whatever.

Cavity wall insulation stands in the way of that heat which used to pass through the inner brick leaf, over the air gap, through the outer leaf and away on the wind. Now most of the warmth passes through the inner leaf, and is stopped. Consequently the temperature of the inner leaf gradually builds up until it is approaching room temperature. This is both good and open to question. It is good because it is saving heat, also because once the wall is warm it begins to act as a low temperature radiant panel, giving back to the room in acceptable form that heat which it has absorbed.

It is open to question because it has a built-in time lag. The inner wall or leaf weighs several tons and will soak up a lot of heat, which during the warming period is not available for the main job of warming the room.

If heating is continuous over several days at least then the incidence of that amount of heat in the total output diminishes. But if the heating is on for long enough to warm the wall, then shut off for quite a time, the warmth in the wall will gradually dissipate into the room and be lost in air changes and the like.

Perhaps it will be clear, then, that in spite of our gratitude to cavity wall insulation, there are times when we would like to exclude even the inner leaf from our heating system. Such times are particularly bound up with speed of response, and this brings us back to air heating. The way to bring insulation right into the room is by the use of a thermal inner lining. This in its most elementary form may be seen in expanded polystyrene tiles, about 3mm ($^1/_8$ in) thick, which can be stuck to the walls. They do a very useful job, and one should always buy a fireproofed grade.

A full scale inner lining is usually a little more thermally resistant than that, also more mechanically strong. It need not make more than one inch difference to wall projection, and a good way to construct one is to fix 15mm ($^5/_8$ in) or 25 mm (1 in) battens to the wall and then clad the wall in hardboard, with mineral wool packing in the gap. The wall must not be subject to dampness, though any such wall may be lined if damp is first totally excluded by a heavy and continuous coating of a bitumen compound or other wholly waterproof substance. The hardboard may then be treated as the wall surface, and papered or painted.

Tests carried out by a firm now unfortunately defunct showed that a warm air system could be shut off overnight, the room temperature falling only a few degrees by morning, and within 15 minutes of starting the room temperature was back to its prescribed level. The overnight maintenance of temperature could be due in part to the fact that the furniture in the room had acted as the heat buffer or reservoir. It was a remarkably good demonstration of comfort with economy.

5 Air Heaters

Enough detail was given in the last chapter to cover the subject of free flow air heaters. They have a place in the total heating scene, in providing a source of maximum heat required while leaving its distribution to nature. In return for this elementary service, which often provides remarkably satisfactory results, is a lower first cost than for a more comprehensive system. It will continue to find support, but this book is mainly concerned with systems which are controllable in greater detail.

The majority of air heaters are gas fired, and an estimated million are installed in this country. They take two forms, floor standing and wall hung. Unlike boilers they do not require or use cast iron heat exchangers, since they do not have to resist water action. Although the air heater does not supply domestic hot water it is now usual for a small and independently operated water heater to be included, possibly as an optional item, within the case. It will share the fuel supply, flue etc.

Both the size and the independence of the water heater are very satisfactory features. It means that the water heater is just right for the job it has to do, unlike a large boiler which in summer is expected to work at low efficiency simply to supply hot water. The independence of both units means that water heating is not dependent upon space heating. More, it allows the air heater to be used for an opposite purpose, for circulating a stream of cooling air in summer, simply by allowing the fan to work while the burner is shut off. In this connection, it is quite common practice for the control system to include a reverse acting thermostat, which will not allow the fan to start until the heat exchanger is up to temperature. In order to promote ventilation without heating this device must be by-passed by means of a ventilating switch, which must of course be returned to 'off' before heating is required.

Another device which is standard equipment on most air heaters is an air filter. Air is no less surprising than water in its capacity for carrying impurities. Much of what is carried in air is invisible to the eye until it accumulates in bulk on a filter. But it is important both to keep the filter in place and to keep it clean.

There are two types of air borne matter which concern us. One is partly what we would call dust, inorganic matter, i.e. ash from a coal fire, plaster from a wall, etc. and partly organic but not living, e.g. lint, minute particles from textiles, plants and so on. The second class of substance is living matter, air borne bacteria and larger organisms. Both classes have a tendency to stick to the hot surface of a heat exchanger, the second class forming a plastic substance which adheres firmly to the surface. The worst result of this is to lower the efficiency of heat exchange progressively. A second result can be a slight tang in the ejected air from the 'cooked' living organisms. While the mechanism is not fully explained, any dust which passes through after heating seems to have an enhanced property for creating a dry feeling on the mucous membrane. Filtration of the air is therefore very desirable.

Cleaning the filter is necessary in order to keep the plant at work: a choked filter can shut it down. A device known as a filter flag is often used. It is a pressure operated indicator, which it is possible to connect to a light or buzzer, which shows when the resistance of the filter has reached a point demanding cleaning. Filters are of two types, washable or throwaway. Instructions are given for care of washable filters, and generally these, after washing, are shaken to remove excess water but put back in a wet condition. Attempts to dry them might cause damage.

All units of this type include the air fan which creates air flow over the heat exchanger and supplies the pressure necessary to cause flow through the duct system. This may lead to an arrangement which is up flow, down flow or even cross flow, referring to the direction of air flow across the heat exchanger. The purchaser cannot usefully become involved in such technicalities, all units of an approved type being capable of achieving a given standard of working efficiency. It might however be relevant to duct connections, and so be preferable for the air outlet to be above, or below, or at the side, in a given case.

It is worth enquiring about the extent and the needs of servicing. Time is money, and ready accessibility saves time. Such features as a bolt-on cover plate which comes away with burner complete, or with instruments, are worth more than a unit which has to be taken apart item by item. Another feature which must be kept in mind is the space needed. A unit may be fitted in a cupboard or under stairs, suitably lined with insulating material but leaving little room to spare. With this in mind, units are often designed to be dismantled from the front, and it is therefore necessary that the unit shall face the openable door, so that with the door open there is ample space in front of the unit. But it should also be ascertained whether side or back access would be required, which could only result in the complete removal of the unit every time it had to receive attention.

Although the pressures employed in a fan powered duct system are very low, it is necessary to take great care with jointing of the flow duct runs generally, and where both the flow and return ducts join the heater. Joints are usually of the socket/spigot type and after putting together are taped. Leakage on the flow side is wasteful, but leakage on the return side, around the heater, can be dangerous, since fan suction can deny combustion air to the burner. If a slide-in filter is fitted, it should be pushed firmly home after cleaning, to leave no air gaps.

Quiet operation has always been one of the makers' aims, and in its way this becomes more important with air heaters than with boilers, because if a unit is noisy its noise can very easily be carried along the large ducts. Although a burner is capable of being noisy this is not often the real worry. It is the fan which has received most attention, and a look across the range of models now on offer shows that makers have abandoned the small fan working with frantic haste, in favour of the much larger, lazier fan, which works at lower speed to achieve the same result. This brings other benefits, such as lower wear, better pattern of air flow over the heat exchanger, and slower fouling rate on the fan itself. (Dust collects on fan blades and must be cleaned at service time.)

Figure 5.1 shows a typical floor standing air heater, the Sugg 25/30M (7.34 to 8.8 kW). It is a down flow unit, the fan being fitted to a slide-out plate in the top compartment. The tall centre compartment houses the burner equipment and heat exchanger, and if required the small water heater which has a branch off the gas supply and one back into the air heater flue. The plenum base, though optional, is necessary, and is not supplied standard in this case because the unit might act as a replacement on an existing installation. This unit, concentrating upon width reduction, measures approximately 1200 mm (48 in) high (without plenum base) by 600 mm (24 in) deep, by 300 mm (12 in) wide.

Figure 5.2 shows an upflow air heater which is designed to accept as much optional equipment as is desired, up to the point at which it qualifies for the makers' description of a Total Comfort System. The illustrations show it acting as a simple heater, air flow entering from below: then as a unit complete with cooling coil, humidifier and electronic air cleaner. Any of the extras may be applied independently.

This unit has a heat output of 15.8 to 45 kW (54 000 to 154 000 Btu/h) and is wholly accessible from the front.

It will be seen that air entry, in the right hand illustration, is not from below but at the side. This is the method which has to be adopted when the unit is situated in a cellar, or on a ground floor of solid construction. It is comparable with the type of unit long known as a basement heater, in which the external connections to both flow and return air are at the top of the unit. Upgrading to a comfort system is of course a special function of the unit shown here.

A unit fitted in a basement or cellar, whether it be an air heater or a boiler, can present a special kind of problem which

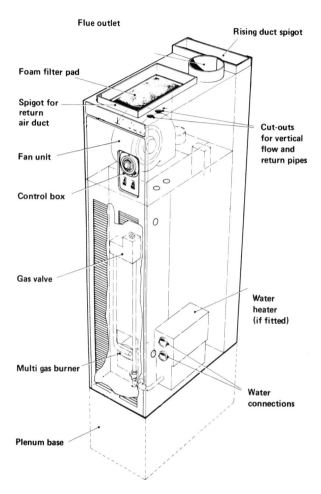

Flue outlet

Rising duct spigot

Foam filter pad

Spigot for
return
air duct

Cut-outs
for vertical
flow and
return pipes

Fan unit

Control box

Gas valve

Water
heater
(if fitted)

Multi gas burner

Water
connections

Plenum base

*Fig. 5.1. An example of a downflow air heater unit with optional
water heater (by courtesy of Thorn Heating Ltd)*

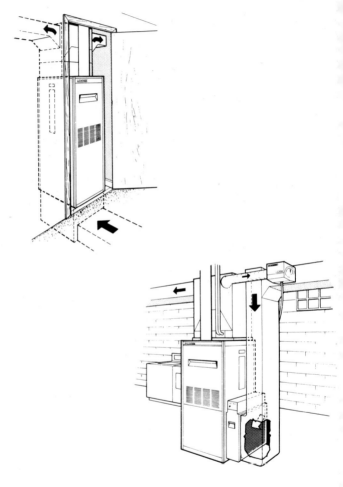

Fig. 5.2. A typical arrangement of a wall mounted upflow air heater (by courtesy of Lennox Industries Ltd)

deserves mention. It must be of the conventional flue pattern (except in special cases of extended duct balanced flue, which is rare) and so the burner draws its air from around the unit. A basement does not have a natural air supply in the way that above-ground rooms have, but the basic minimum need (see Chapter 6) is no less important. It should not be left to chance, to the likelihood that enough air will find its way down via the stairs, even through a shut door. The way to handle this is to run a vertical duct, of 100 mm (4 in) diameter or greater depending upon the heater. The upper end of the duct should be well clear of ground level, and fitted with a mesh cover. The lower end should terminate in the vicinity of the air inlet to the burner.

The free standing and wall mounted units broadly described so far are typical products of those firms which are members of the Warm Air Group of SBGI, the Society of British Gas Industries. Enquiry at any gas showrooms will produce more details of such units. It should be noted that each manufacturer makes a range of heaters, in size of rated output, and almost always offers some variation in physical size and shape, to fit varying installation conditions. For that same reason there will be a choice of connection positions, for air flow and return, perhaps for water connections to a water heater.

In many cases what are structurally the same heaters may be equipped for oil firing instead of gas, and such details may be obtained from the major oil companies or their local agents.

We will not take a great deal of space over units which are thoroughly catered for in readily accessible quarters. But perhaps we can devote more space to what might be called minority types, simply because they are less well known. There are two which deserve mention.

There is Afos, a combined air and water heater. It has two sections, boiler and air heater, usually allied within one casing but capable of being fitted separately. After the burner, which may be gas or oil, the heating medium is water, and a generous supply of domestic hot water is said to be available. The heat exchanger in the air heater differs from that in a direct fired

unit, in being at much lower surface temperature. Consequently
it avoids most of the charring of dust and microbe particles
mentioned earlier. Heat into air is controlled by modulation
with room thermostat operation, and this is a feature which

*Fig. 5.3. An example of a combined air and water heater
(by courtesy of Afos Ltd)*

compares well with the more usual on/off mode of operation.
The fan, protected by a washable filter, is appropriate to a full
duct system, and in addition to having variable fan speed
control the unit may be given a programmer.

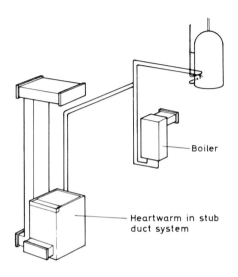

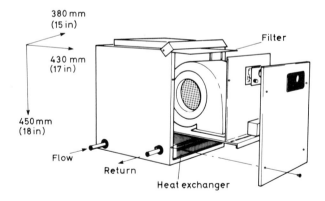

Boiler

Heartwarm in stub
duct system

380 mm
(15 in)

430 mm
(17 in)

450 mm
(18 in)

Filter

Flow

Return

Heat exchanger

*Fig. 5.4. The Heartwarm air and water heater system in a stub
duct system (upper) (by courtesy of Massrealm Ltd)*

The other type of unit relies upon the existence of an independent boiler, and like the Afos it warms air in a water-to-air heat exchanger, with the same physiological benefit. The unit is the Heartwarm of Massrealm Ltd. In dimensions it is an exact fit as a replacement unit into one of the original Sugg air heaters (coming from the same source) and is therefore tailor made to avoid having to make duct modifications. But more than that it is quite suitable to act as a unit in its own right, with enough fan power for a full duct system.

As a replacement unit it assumes that the old burner and heat exchanger are both in need of replacement – hence a new boiler, and a heat exchange section complete which takes far less room than that which it replaces. As a new unit it emphasises the principle that a heater needs a flue, and a warm air generator needs a good position relative to duct work, and the two are not necessarily coincidental. With Heartwarm each may be given its own best situation, and simply connected by hot water pipework. The system may be regarded as a normal wet system in which the entire heating load is concentrated at one point, in the heat exchanger of the Heartwarm. Domestic water will be available in the usual way. In some circumstances it may be claimed that this is a variant of a wet system with no long circuits capable of suffering damage, or frost when idle.

Electric heating by warm air systems is dealt with in Chapter 9. The principal unit concerned with duct systems is Electricaire, the large storage heater working on off-peak current. Individual storage radiators achieve much the same effect but not centrally and by duct.

The use of electricity at standard tariff to provide warm air, by oil filled radiator or by fan convector, cannot be included in a list of standard recommended methods. It is bound to be expensive and if adopted should be done in full awareness that the basis is convenience, not economy. The weekend cottage might be a suitable place for such a system, where the 'on tap' element is worth a lot.

Solid fuel is no longer a serious contender in this market, except in the following respects. Room heaters have a substantial part of their output in the form of warm air rising off the appliance, and kept in the room by strict control of the chimney opening. This is a valuable part of the total efficiency, but it is, as it were, freelance, not a part of a system. A solid fuel boiler may be used to supply the hot water fed to the Heartwarm unit just described, provided that the system includes a hot water cylinder or other buffer for excess heat.

It was not unknown for a solid fuel stove to form the heat generator in a brick central unit, and so long as all the safety precautions against fire and overheating are observed this remains valid. But the decline in brick centrals has almost eliminated that type of heat source.

6 Flues

All the so-called fossil fuels, plus other substances such as wood and wood products, which contain carbon, burn to give off gases, the products of combustion. Our ancestors, and we are led to believe the North American Indian, mastered this elementary fact when they left a hole in the top of the dwelling so that the smoke could escape.

Nowadays we have flues of varying degrees of complexity, which may be broadly classified as conventional and other. The conventional flue is in most cases the traditional chimney, an integral part of the house structure, a tube conveying waste gases from the fireplace to the open air above. It works because the column of warmed gases inside is lighter than an equivalent column of colder air outside, and the lighter column rises. But this is a delicate balance, and easily upset. A down draught caused by some vagary of the wind will make the chimney 'smoke'.

Some chimneys smoke continually, or always when the wind is in a certain direction. The first may be due to a structural fault or to a permanent adverse condition. The second can sometimes be corrected by neutralising the wind effect. A high building, tree or hill will sometimes cause a wind to lift then to plunge, unfortunately, on to the chimney. There are various types of cowl which are claimed to remedy this or other fault, and would-be purchasers would apply a worthwhile safeguard if they were to insist upon a cure-or-money-back guarantee in writing before buying.

It should be evident that a good chimney will not have a lot of bends. It will, if of any age, be far bigger than is necessary, and the worst offender in this respect is the much praised ingle-nook type of chimney, with ample room in it to accommodate the chimney sweep. Such a chimney is disastrous in modern conditions and almost always smokes. To examine the reason it does so will outline some principles which can be more widely applied.

The ordinary chimney, the nominal 230 mm (9 in) square one, can evacuate up to ¼ *ton* of air per hour. Imagine the appetite of the big one! But what is just as bad, it starts at a high level, perhaps 2 m (6 ft) from the floor. That was all right when wood was burned as huge logs. But nowadays it is

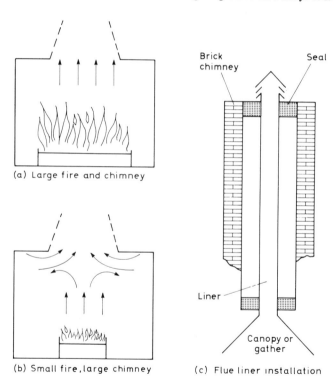

(a) Large fire and chimney

(b) Small fire, large chimney

(c) Flue liner installation

Fig. 6.1. The chimney must be efficient and properly sized for its job. With a large chimney and a large fire (a) it works, but with a small fire and a large chimney (b) the balance is wrong. This can be improved by using a flue liner (c) and a canopy or 'gather'

more often expected to cope with a small basket grate or something on that scale. If we assume that in both cases the chimney is at work, having a rising current in it, then the difference in the two cases can be seen in *Figure 6.1*.

In the first case the fire is large, its waste gases voluminous and over a large area. They monopolise the entry to the chimney base. In the second case the chimney, as always, taking what is most ready to hand, satisfies its appetite with air from the neighbourhood of the entry. Being satisfied, it turns away the legitimate exhaust, the smoke from the fire. Hence, the chimney smokes.

Part of the solution is to bring the entry nearer to the fire. The only way in which this can be done with such a chimney is by means of a canopy, an artificial extension. But a canopy or extension of the same cross sectional area might fail because it is too large, pulling in diluting air from the sides and so failing in the middle. So there has to be a reduction in area as well. The easy modern way to achieve this is by having a flue liner. This is in effect a new flue which uses the run of the old one for convenience. The old one must be rendered ineffective by sealing off the annulus between old and new. A flue liner can have several beneficial effects.

(1) It concentrates the flue upon the job in hand, leaving no margin for adventitious air evacuation.

(2) It bypasses any troubles due to deteriorating joints in chimney brickwork.

(3) It evacuates flue products at higher velocity, giving more positive results, less condensation in the flue, better carry-off of flue products from the terminal. A higher exit velocity will often overcome a tendency to down draught.

These comments upon the use of canopies etc. were designed to show the difference between a suitable modern chimney and an unsuitable one, and were not strictly relevant to the matter in hand. But it allows us to point out that all modern appliances of the types we are considering require to be connected directly and positively to a chimney or flue, and not left with an open ended pipe discharging in the general direction of a chimney base. That fact underlines the sense and logic of a flue liner, which avoids great changes in cross sectional area as the flue gases go from appliance to terminal.

The appliances themselves have stub flue pipes of the correct diameter, pointing either upward or backward, usually depending upon whether they are intended to stand against but outside of the hearth or chimney wall, or to be at least partly let into the hearth so that an upward flue pipe points in the right direction.

Boilers which incorporate a flue stabiliser have upward pointing stub flues for that reason, and it is expected that to couple into a flue behind, a bend will be used. When flue runs are given a change of direction it is preferable to use slow bends, of the order of $45°$, and not right angles.

Gas fires often have a closure plate, a sheet of metal or other material which shields the fireplace opening, and has two holes in it; one for the flue to pass through, and one at or near the bottom which acts as a static draught stabiliser. The closure plate must be firmly secured to the hearth, to avoid unsafe conditions arising. But the gas fire/back boiler, which concerns us, has the gas fire flue transferred backwards to a common flue manifold over the boiler, the lot finishing in a spigot designed to fit straight into a flue liner.

If an appliance is fitted so that a backward facing flue enters a larger brickwork chimney at right angles, care must be taken to see that the stub flue is correctly set. As *Figure 6.2* shows it should project a little beyond the inner near wall, but still leave the greatest distance from the back wall. Failure to attend to this, for instance in the case of a pressure jet oil fired boiler, can result in a very disturbing amount of muffled reverberation, as the jet cuts in and during working.

When flue pipes pass through walls, plates and the like, and have to be sealed in, due allowance must be made for the frequent expansion and contraction which takes place. It is useless to make the joint with cement, even with fire cement or fireclay. A soft heatproof jointing must be used, and the most obvious one is asbestos string or asbestos rope depending

upon the size of the annulus. If there is any objection to the
use of unshielded asbestos then it may be shielded by a collar
which is attached to the flue pipe but not to the back brickwork
or metal.

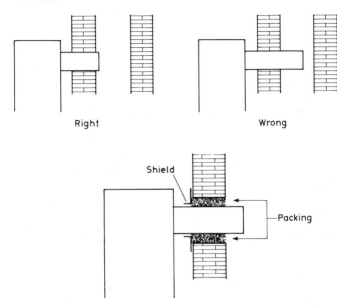

Right   Wrong

Fig. 6.2. Installation of an appliance with a stub flue into the chimney

Damp patches on bedroom walls are often the result of
condensation inside a flue, itself resulting from slow lazy flue
gas travel, and the condensation penetrating bad brick joints.
A flue liner cures this.

A flue terminal should always be fitted to a flue liner, and
advice upon both is obtainable from the representative of the
fuel to be used (Gas Region etc.).

The chimney with an open fire, the culprit which can take
away ¼ ton of air every hour, was in some respects better
behaved than a more modern combination would be. The
effect of all that air was to dilute the flue gases, often to the

point at which condensation would not take place. But now, consider a gas or oil fired boiler, or modern solid fuel appliance, with a flue spigot of correctly designed size. In most gas boilers for instance this is 100 mm (4 in) in diameter, giving a cross sectional area of about 78 cm^2 (12.5 in^2). That is the size the flue should be, all the way up. If the spigot discharges through an aperture in a blanking plate into a conventional 230 mm (9 in) square flue, area 529 cm^2 (81 in^2), it has more than six times the area it needs and will dawdle. Worse, because of the blanking plate there will be no huge volume of excess air to dilute the flue gases, and condensation will be very much more likely to occur. The evidence all points therefore to the advantages of having a flue liner fitted in an existing flue or chimney to go with a modern appliance.

Because of the way in which wind is affected by high obstructions, it should always be the aim to carry the flue terminal above the roof ridge by about 0.6 m (2 ft). Any wind striking the ridge will then be rising at that point, which is beneficial.

Having more chimneys than are needed, as in older houses with one open fire per room, is undesirable because of their property of whisking away air, often air which has been expensively warmed. The permanent way to deal with them is to cap them off at the top with a waterproof and windproof cover. Then, just in case the exposed brickwork is porous, do not wholly seal the lower or fireplace end. Instead, perhaps after removing the old grate and surround, fit a sealing plate but leave in it a slot, mesh covered and no more than 150 x 25 mm (6 x 1 in). This will prevent stagnation. To remove the breast and associated brickwork is builders' work and should never be undertaken without full professional appraisal of the consequences to the rest of the structure.

At the other extreme is the home without a chimney. It may be a mews conversion, or a new house built with wholly electrical heating in mind. In most cases a conventional flue can be added. One way is to use asbestos flue pipe, double walled to keep condensation to a minimum, with a terminal. Such a flue may be run indoors, in which case there are strict provisions in the Building Regulations about the path it will take. In particular it must be kept away from structural timbers. Alternatively it may be run mainly outdoors, where the asbestos double wall really comes into its own.

The easiest way to run an outdoor flue and still conform with the need to have the terminal above ridge level is to run it up the gable end, other things being suitable. In that way it may be supported off the wall for most of its run.

A more robust, and in most cases better looking job is made by the use of proprietary precast flues such as TrueFlue make. These are in the main consciously styled, have all the necessary insulation built in, and do not need the same degree of support.

The best advice that can be given is: if you have a good working chimney, use it, in preference to any of the other methods of flue gas removal mentioned here. When coupling a modern appliance to a traditional chimney, follow the instructions given by the makers of the appliance. Some measure of sealing in will almost always be required.

There are of course nowadays types of dwelling in which even one chimney per dwelling would be quite out of the question. Multistorey blocks of flats are an example. If these

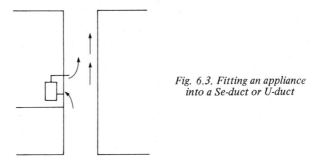

Fig. 6.3. Fitting an appliance into a Se-duct or U-duct

are to be used for other than electrical (unflued) appliances, then one way is to incorporate a Se-duct or U-duct. We need not give much consideration to these here. They are large centrally placed flues into which appliances from adjacent flats pass their combustion products and from which they draw combustion air. Se-ducts and U-ducts are an integral part of the main structure. Unlike the domestic chimney they cannot be added on at any time, but they are not domestic items except in the collective sense, as in flats.

The balanced flue

This is applicable mainly to gas fired appliances, occasionally to oil fired ones, never to solid fuel. The balanced flue forfeits dependence upon the traction effect of a chimney in return for the assurance that any disturbing influences acting upon the flue outlet will act equally upon the air inlet and so cancel out (*Figure 6.4*). This is achieved by ducting the flue outlet and air inlet to a common terminal. Then, if we suppose that a head-on wind is blowing, which would cause an ordinary chimney to reverse, its effect here is counteracted by the same force blowing upon the air inlet and so through the system. It is of course a condition that the combustion chamber must be sealed, hence appliances with a balanced flue are often called room sealed.

The theory of the balanced flue is as follows: suppose that the discharge pressure is p and the suction is s, then the differential causing flue operation is $p + s$. Now suppose a wind of force P blows on the terminal. The discharge pressure becomes $p - P$ and the suction becomes $s + P$. The differential is $(p - P) + (s + P) = p + s$, i.e. it is unchanged by P.

Balanced flue appliances are usually situated against an outside wall, though some will if required work with extended ducts between appliance and terminal. There are a few rules about the placing of balanced flue terminals, social and technical, which must be observed. A terminal must not be placed near an openable door or window, or at low level near a public right of way. It must be covered by mesh or similar to keep out intrusive objects. For good operation it must be kept away from building corners, and from too near bushes or other buildings. Such features can cause non-homogeneous wind effects.

Traditional flues, Se-ducts, U-ducts and balanced flues have one thing in common. They work for ever for nothing, being dependent upon natural phenomena. Such items as the balanced flue have gone a long way to destroying the

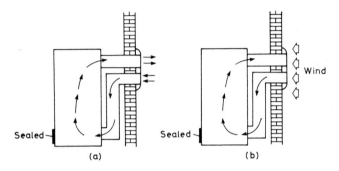

Fig. 6.4. How a balanced flue works

monopoly of the hearth as the focal point of the room. But there are still cases in which it may be impossible or undesirable to put an appliance anywhere suitable for connection to a natural flue. It must not be supposed that there is yet a universal answer to that, but in some cases, to be ascertained by enquiry, an appliance may work with an unnatural flue, e.g. a long horizontal run, to which the power is applied artificially. Such flues are 'assisted'. They include quite complicated controls for safe working, which ensure that (1) the fan starts before the burner lights, and stops after a purging period when it goes out; and (2) if the fan fails to start the burner cannot light.

There is, clearly, a linking of controls between appliance and flue which makes it necessary for the appliance maker to be involved. It scarcely ever happens that a domestic situation cannot be adapted to match a natural flue, but anyone who believes that his does not should consult the gas authorities, or the representative of a major oil company.

A much simpler and cruder device may be used with solid fuel appliances attached to a chimney which is reluctant to work. This is a propellor type fan, let into the chimney, often at an angle, which assists the upward movement of air in the chimney and so creates draught. It carries no special conditions,

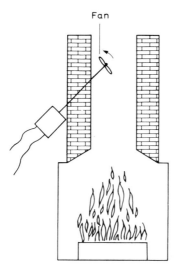

Fan

Fig. 6.5. A fan assisted flue for a
solid fuel appliance

may be switched on or off at any time and without regard to the state, or even the existence, of a fire. Its greatest benefit is probably in cases of marginal or intermittently bad draught. A fuel supplier, particularly if accredited by the SFAS (Solid Fuel Advisory Service) should be able to advise.

Care of flues and chimneys

With the coming of the Clean Air Acts domestic chimneys no longer carry and discharge such volumes of heavy smoke, loading the walls with soot. One must however expect a certain amount of soot to collect, from solid fuels and from oil burning. In addition solid fuels may give rise to mineral dust (fly ash) which will rise part way up the flue. Another reason for occasionally inspecting a traditional chimney is that solid materials, parging and the like, have a tendency to come adrift and fall to the bottom. In time these could build up and choke the applicance stub flue, leading to a dangerous condition. Therefore, chimneys which used to be cleaned every six months should now be at least inspected once a year. The most that a balanced flue terminal on gas needs is occasionally to make sure that the outside of the terminal is clear. Little boys do try to 'post' sticks and things into any such opening, hence the terminal guard. A balanced flue system used with an oil fired appliance should be examined for soot formation during the routine servicing of the appliance.

Assisted flues of the complicated type should be looked at to see that all joints are tight, all controls working as they should (by simulating the conditions which make them work), and that flue runs have not sagged and developed collecting points for condensation. This is best done by experts.

Ventilation

It is most appropriate when considering flues, which take away products of combustion, to consider also the rules which deal with the supply of air to the burner – without which there would be no products to take away.

The most thorough investigation into all aspects of ventilation we owe to the gas industry, and them we will quote. But the principles, and in the main the figures, can be applied to other combustion processes.

Perhaps as a reaction against generations of suffering cruel draughts, people have become very draught conscious. This has led to many cases of over-reacting, and draught stopping to a total extent. Such operations have too often ended in the coroner's court. The gas industry, whose products of combustion are invisible, unlike the recognisable vapours from bituminous solid fuels, is particularly vulnerable to such happenings and so is most concerned by them. They have therefore specified the minimum requirements for the areas of permanent free openings into any room in which an appliance is at work. A full list is available at the gas showroom, but the main points are given below.

For boilers and similar central heating units with conventional flue the *minimum* opening at *high* level into the next room must be 6.5 cm^2 (1 in^2) for every 0.58 kW (2 000 Btu/h) of maximum rated output. At *low* level the minimum is 6.5 cm^2 (1 in^2) for every 0.29 kW (1 000 Btu/h). Half these areas are needed if the openings connect directly to outdoors.

Room sealed models require the same areas at high level, but at low level only half the allowance for conventional flue models. Room sealed appliances are those having a balanced flue or Se-duct.

The minimum areas given above are permanent free areas, i.e. in a typical grille they are the *openings*, not the total face area. Although the proportion of permanent opening varies from grille to grille, it would be unwise to assume that the area of opening is more than 30% of the total, and it would be wise to choose generously sized grilles.

The principles, and to a large extent the details, given by the gas industry may be held to apply to any fossil fuel which is burned under controlled conditions (this excludes the open solid fuel fire). If in doubt be generous, for it is most necessary that air shall be freely available to the combustion process. If the flame is starved of oxygen there is trouble ahead. Most of the complaints about open fires smoking arise through air starvation (brought on by draught proofing) which is caused by the phenomenal appetite of that appliance for excess air. It is an unbeatable spiral, showing how unsuitable uncontrolled combustion processes are in the modern home.

With the accent so much on supplying air for combustion, it might be asked why room sealed or balanced flue appliances require air openings. The answer is that they are what they claim to be, namely for ventilation. Combustion appliances emit a certain amount of warmth, and unless this is removed by deliberate ventilation it can build up a local temperature.

This could become uncomfortable, in some circumstances even dangerous.

Ventilation control

The operative word is control. A lot of houses can be shown to leak like baskets, with air movement taking place between floor boards, through window and door jambs, up unused chimneys, from room to room and away in the attic, and so on. A serious effort should be made to stop all of these, because they carry away a lot of the air so expensively warmed by the heating system. Only when stray and unofficial draughts have been stopped can we realistically set about controlling air movement. Control has two meanings. There is the air allowance for the room containing the fuel burning appliance, while for other rooms it is generally accepted that there shall be 1½ air changes an hour, not least to avoid building up a concentration of carbon dioxide in occupied rooms.

The stopping of unregulated draughts is usually thought of as a part of an insulation programme, and we will treat it as such and discuss it in Chapter 10.

The deliberate provision of air movement we will look at in two ways. First there is the air for combustion: if it is admitted, for example via a door, and the appliance is against the opposite wall, there is bound to be a draught cutting across the room. But if the air opening is on the same wall, even adjacent, then the body of the room will experience no direct draught. A way to achieve this may be seen in the Baxi type of fire with an air duct from outdoors, under the floor, emerging beneath the grate (*Figure 6.6*). But it could, by travelling under the floor or on the floor, discharge near to the fire, just as readily. An opening made in a wall for such a purpose rarely needs more protection than that which excludes wild life. Although a flap may be fitted, the fire when out will not exert any pull on the duct since it is making no demand for air.

If this simple scheme is not possible, or not welcomed, a ground floor room with suspended floor might even benefit by having a mesh covered air opening cut into the floor. So long as the floor is well covered, for instance using good quality under-felt, it will not noticeably increase heat loss. But it will ensure the ventilation under the floor which is necessary to combat dry rot.

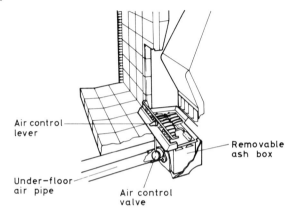

Air control lever

Removable ash box

Under-floor air pipe

Air control valve

Fig. 6.6. A typical underfloor draught fire

If in the end we come back to the door as the source of fresh air, do not rely upon air entering under the door. It not only gives the worst draught, it is also possible for someone else to stop it off, for example by fitting a thicker carpet. The proper way to use the door is to have a deliberate opening at the top, perhaps by cutting an inch off the top.

Or one might borrow a trick from warm air practice, and let a ventilator into a wall which communicates with another part of the house. This is the kind of detail which ought to be decided on personal grounds, for example is it preferable to encourage air movement to or from a room other than that to which the door leads, or should air movement take place in some other than the door-to-fireplace line?

A special example, open to be borrowed for special circum-stances, comes from warm air heating of the 'natural convection' type. It involves fitting an extractor fan at high level in a wall, in order to move warm air from one room to another or to a hall for onward travel.

It is difficult to think of any case in which the last example would be required. The general rule is that combustion processes will grab all the air they need, so long as the air is free to travel their way. This means leaving ample area of grille etc., but in one connection there must be a warning. It is by no means unusual to have an extractor fan fitted in the kitchen, with the sole purpose of removing moisture laden air from the kitchen. Very often the boiler is in the kitchen too. In that case the air inlet allowance of 6.5 cm^2 per 0.29 kW (1 in^2 per 1 000 Btu/h) is inadequate, since it has to supply air for both boiler and extractor, and the latter is more powerful. There is then no rule about the size of opening, but it must be judged according to the results of a 'spillage test' which an expert can carry out. This involves putting everything to work, including the fan, and then finding out whether air travels continuously forward, or whether there is any tendency for flue gases to 'spill' back into the room.

Finally, we cannot put too much stress upon the need for air openings to be *permanent*. They must not be closeable at any-one's whim or for any reason.

7 Fuels

The fuels used in domestic practice may be listed as:

Solid fuel: coal (including anthracite); coke; wood; peat. The last two are rare in the UK and we will not pursue them in any detail.

Oil: two grades, kerosene and gas oil (also known by other names).

Gas: mains distributed gas is, except in some remote regions, now entirely natural. Bottled gas, as propane or butane or a mixture is readily available.

Electricity: now almost entirely standardised in characteristics, but sold under a variety of tariffs.

One of the more impressive facts about the technological evolution of domestic heating appliances is that the combustion apparatus is no longer a sort of indoor incinerator, a place to burn any old rubbish. True, an open fire will still burn nut shells if required, but its principal fuel must be selected with care, both for technical and for legal reasons. The same is true of all oil and solid fuel appliances, where there exists a choice.

The manufacturer of an appliance will name suitable fuels for use with his appliance, and the choice should be made from such a list, not only as to type of fuel but also to size and grade where applicable.

On such detail depends the efficient performance of the appliance, and in certain cases one's immunity from prosecution under the Clean Air Acts. Let us look a little more closely at the choice of fuels.

Solid Fuel

Bituminous coal is familiar to most people as 'house coal', whose unregulated use was a considerable factor in causing smokeless zones to be declared. Any appliance which sets out to burn bituminous coal in a smoke control area must be specially designed to be what has become well known as 'a smoke eater' and must have gained exemption under the Clean Air Act 1956. Bituminous coal must not be used in boilers, where it would deposit soot on the heat exchanger surfaces and so ruin the efficiency of heat exchange. It would indeed be fair to say that bituminous or house coal, now far from being the commonest of all solid fuels, is hedged in with special conditions which greatly restrict its use.

Two ranges of smokeless solid fuels have grown up. One consists of naturally occurring fuels, the other of manufactured fuels which are derived from coals in the bituminous range by removal of most or all of their smoke producing constituents.

All these fuels are dearer per ton than house coal but not necessarily dearer per unit of heat output since they lend themselves to more efficient use. This in turn warrants the design of more efficient appliances, and sharpens the attention to be paid to correct size or grade of fuel, upon which the best operation of the combustion process depends.

The natural smokeless fuels are anthracite and Welsh dry steam coal. Anthracite is a relatively pure form of carbon, with very little ash or volatile matter. Because of the latter it is not easy to light, and anthracite burning appliances are often equipped with a gas poker or other device with a similar purpose.

The common sizes of anthracite, in descending order, are: French Nuts; Stove Nuts; Stovesse; Beans; Peas; Grains. The last two, occasionally the last three, have a fan assisted fire bed to overcome the denseness of packing due to the small size of the fuel. Such a fan performs a secondary role as a combustion controller. When the fan is off the fire will idle.

Welsh dry steam coal has a low enough volatile content to be acceptable under the Clean Air Acts without special dispensation, but not so low that it is difficult to light. It is customarily supplied in the larger size range, as Large and Small Nuts. It is incidentally known variously as Welsh Nuts, Welsh Steam, Dry Steam, or Welsh Boiler Nuts.

There are three classes of prepared or manufactured smoke-less fuels. These are: coke, semi-coke and briquettes. Since the passing of gasworks there is no more 'soft' or gasworks coke, and the only coke now available appears under the name Sunbrite, in sizes called Singles, Doubles and Trebles. The appliance manufacturer will state which size you need. Some assistance with initial lighting of a Sunbrite fire bed is desirable.

The semi-cokes, familiarly known as 'solid smokeless fuels', the best known being Coalite and Rexco, are the result of giving raw coal only a part of the treatment which would yield a coke. They therefore retain enough of the original volatile matter to assist ignition and to give some flame movement in a visible fire; but not enough to make them offensive under the Act. Both Coalite and Rexco bear a resemblance to the original coal substance and are sold in the larger sizes mainly.

The third class, briquettes, are reconstituted into a con-venient size and usually an ovoid shape. Phurnacite and Home-fire are both made by the National Coal Board. They are made from selected grades of coal ground very fine, and passed through a press to give the form required. Phurnacite briquettes include a proportion of pitch added as a binder, while Home-fire starts with a coal of higher volatile content and relies upon heat generated by the press to employ some of the inherent tar substances as a binder.

There is a further difference, that Phurnacite is carbonised, i.e. heat treated to drive off excess volatile matter (including that which was contained in the binder) *after* being briquetted. But Homefire is made from a char, which is coal substance already partly devolatilised in a fluidised bed. Anyone interested in pursuing these matters in a not-too-technical way will find a clear presentation in a book called *Solid Fuel in the Home*, published by the Women's Solid Fuel Council from Hobart House, SW1X 7AE.

For the average reader it will probably be sufficient to know that Phurnacite is very well suited to solid fuel cookers, room heaters and independent boilers. Roomheat is suited to modern open fires and to glass fronted and other types of room heater.

Any discussion of solid fuel must touch upon storage. Solid fuel of all kinds should be stored in the dry, the importance of this increasing as the surface porosity of the fuel increases. Thus, most of the rain will run off the hard surfaces of natural fuels of all but the very small size. But coke and semi-coke will be able to absorb a very high proportion of its own weight as water, which will have to be driven off by the fire bed when it is charged, with considerable loss of efficiency.

Then we come to handling. All solid fuels degrade to some extent when handled, and even when under the weight of their own heap. Degradation leads to size reduction, and the production of fines and dust which are usually thrown away – a complete waste of money. Handling, moving from one place to another, shovelling over, even walking on top of the heap, should be avoided. And to minimise the slow degradation at the bottom of the heap it is desirable to draw off fuel from ground level so that the bottom layer does not remain unchanged for a long time.

The tendency to degrade is higher among the prepared fuels than it is in the natural fuels. Thus, long after the chemical and combustion features of briquettes had been settled, scientists were still wrestling with the question of mechanical strength. The principal problems were overcome, of course, but a briquette should still be regarded as needing care in handling and storage.

Solid fuel differs from all of its competitors in that it will not obey instantaneous commands. It cannot be switched on and off with immediate effect as can gas, oil and electricity. This feature brings with it other considerations, some good and some less good.

The kitchen based boiler, the hearth fitted appliance and the combination range, when their dampers have been shut either by thermostat or by hand, continue to 'tick over' at a reduced rate of combustion and when the next command comes to start work the fire picks up slowly to a full working condition. The noticeable feature, living with one of these units, is that it is always comfortably warm in the vicinity of the apparatus, which makes its own rather involuntary contribution to house heating in this way. A secondary facet of the same phenomenon is that even though the fuel may be difficult to ignite, it is a job rarely required, in some cases only at the start of the heating season.

A more limiting aspect of the lag in answering to controls is that there has to be allowance made for it, and particularly when it is being ordered to cool down. Consider a boiler, with thermostatically operated damper, which is working at normal high capacity in a small bore heating system in which the room thermostat operates the circulating pump. The house warms up, the room thermostat becomes satisfied, and stops the pump. No more hot water is going to the heating system. The boiler thermostat quickly senses this, and closes the boiler damper (and stops the forced draught fan if fitted). But the boiler does not stop its heat output. It may even increase it temporarily, but will in any event go on delivering heat in considerable quantity for quite a time. If this heat had nowhere to go outside of the boiler it would have to be absorbed into the quite small quantity of water which the boiler holds, and almost certainly would lead to boiling and in most cases to a state of considerable danger.

In such a situation there must therefore always be a substantial buffer available, a large source of heat transfer permanently connected to the boiler which cannot be isolated from it. This is the domestic hot water system, the cylinder of water containing over 100 litres (or about 25 gallons) which can safely be raised about 40 degF or 20 degC.

Thus it will be clear that a solid fuel boiler cannot power a heating system only, but must include domestic hot water. It shows too that the domestic hot water system must not be fitted with a cylinder thermostat, which could stop the primary flow into the cylinder and so isolate the boiler from the cylinder.

Fuel Oil for Domestic Use

There are two grades of fuel oil sold for domestic heating. They are known traditionally as kerosene and gas oil, but it is not uncommon to find them called 28 sec and 35 sec oil (a measure of their viscosity) or the first as domestic fuel oil. The latest, official system, by which they will be recognised in an approvals document, is as Class C and Class D fuel. We will call them C and D; C being the lighter kerosene and D the gas oil.

They are not fully interchangeable. Some appliances will burn only one of them. Other appliances, which have a ready means of varying a component of the burner, can be adapted to burn either but the adjustment is a job for an expert. Broadly speaking, boilers of the pressure jet burner type will be found using Class D, other appliances Class C. But in the first case there is often scope to adjust to the other fuel. The arrangement quoted does however fit neatly into another aspect of modern thought, namely that pressure jets are not acceptable indoors (because of their noise level) and Class D fuel is not welcomed indoors because, in the event of leakage, or spillage during filter cleaning, etc, its smell has a lingering quality beyond that of Class C. Class D and pressure jets are therefore relegated to an outhouse together.

Fuel oil is almost always fed to the burner by gravity, i.e. the storage tank is at a higher level. If not — if for instance it happens to be below ground — then the oil must be delivered to the appliance by pump. In the case of a pressure jet boiler this can be arranged, by a slight modification to the oil pump which forms part of the burner, plus extra pipework to form a two-pipe system. For any but a pressure jet an independent pump must be fitted, and again a two-pipe system is required. The system operates by the pump taking fuel from and returning it to storage, the burner taking off what it needs at any given time. The system will be explained in detail when necessary by an authorised agent of one of the major oil companies — the one chosen to supply the oil, most likely — or by a competent installer. The general principles will be seen in *Figure 7.1*. It is usual to find that practical limits are assigned to pump performance, which in the case of gravity feed are the maximum and minimum heads. Typical values for these might be 8 and 1 m (24 and 3 ft). The limit at which suction can occur in a lifting system is of the order of 3m (10 ft).

All pipe runs should assist flow by having true falls, and freedom from dips. Great stress is placed upon having a filter of the right type in the line. It is customary nowadays for this to be outdoors where it is more easily cleaned. Whether or not a fire valve is to be fitted as a compulsory feature is a matter for the local authority to decide in terms of their building regulations.

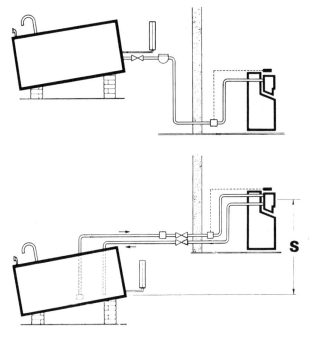

Fig. 7.1. Single-pipe and two-pipe oil supply system to a pressure system jet boiler

The size of the storage tank, and the shape for a given size, are sometimes determined by the limitations of the space available. When that factor is absent, the size of tank is usually balanced between the way tank cost increases with size, and the way oil prices can decrease in relation to the size of each consignment. It is not a matter which requires the services of a chartered accountant on each occasion. Everyone in the trade can now give sensible advice based upon long practice, and only special circumstances should be allowed to overturn this advice. If for example you live on a track which is impassable to lorries for three months each winter then your *minimum* capacity must be that which will last for more than 3 months.

Routine attention starts with cleaning the filter, and opening the drain cock to release any water which (seemingly against all the odds) has collected there. Sludge too may come away, but if what comes is run into a bucket, the good oil after settlement may be returned, so run off generously.

The tank itself will need some maintenance from time to time. Rusting of a steel tank occurs quite readily when paraffin type oils are about, and it is necessary to remove all the rust, dirt and oil, and repaint, using a hard skinned top coat which will resist oil penetration for the greatest length of time. Another point to be noted about oils and metals concerns pipework. Never use galvanised pipe to convey fuel oil. Copper remains best, but any other of the usual metals may be used.

Oil, irrespective of which class, demands quite precise conditions for correct burning. Incorrect burning can lead to smell, sooting of flueways with increasing loss of heat exchange, and general loss of efficiency. Modern appliances are a long way from the old wick burner heating stove, and even that was prone to smoke if the wick was not trimmed.

The best safeguard, and really an essential feature of ownership, is skilled servicing at regular intervals to be agreed with the service engineer. If a service contract can be arranged it saves having to remember and could save money. Beyond that, the owner should be prepared to look occasionally at the appliance, in order to see that operations are normal. If for instance the flame cannot be seen through the inspection window because of soot a more thorough inspection is required.

Gas

Gas is available to all of us, in one form or another. In the form of natural gas it is expected to last until the end of the century, which is long enough for our purposes. With the exception of a very few remote districts, mainly in northern Scotland, manufactured gas is no longer distributed. There remain many places in the UK which are not within economic distance of a gas main, though in total of dwellings they do not amount to much. For them gas can be supplied in containers, principally by Calor.

Manufactured gas, natural gas, Calor gas, are all different, but the differences are really no concern of the public. Two facts alone concern the user or potential user of gas.

(1) The gases are not interchangeable. An appliance, such as a boiler or a cooker, may remain structurally the same for any gas, but it will be given a wholly different set of combustion parts for each.

(2) In order that the preceding point does not cause despair, the whole matter of interchangeability, the right bits for the job, and the overall safety and efficiency of the unit, are very jealously and zealously guarded by the suppliers, either British Gas or the commercial organisations. The proper functioning and safety of a domestic installation is now insisted upon by law.

There is still some noticeable reluctance to use bottled gas as an automatic alternative to a non-existent piped supply. Yet it is almost wholly equivalent, and the price shows no significant difference. In place of a perpetual on-tap situation one may enter into a contract or arrangement with the supplier for regular refills. This is best organised by having two cylinders or containers. The one is used until empty, the supply changed over to the other, and the empty one is replaced by a full one. The only slight chore is to transfer the supply pipe from one to the other cylinder, and rural life offers far worse tasks than that. Natural gas and bottled gas are not lethal in the way that town gas was, since they do not contain carbon monoxide. In heavy enough concentration they can asphyxiate by excluding air, but that takes some doing.

The products of combustion could contain carbon monoxide if combustion is interfered with and not allowed to proceed to its normal conclusion, which merely emphasises the need for ample combustion air, as detailed in Chapter 6, and also the need to keep flue ways unobstructed. But, as has been pointed out, fatalities from this cause are of the order of those due to being struck by lightning; and *we* would point out that reasonable care gives greater chances of avoidance in the case of gas.

In all matters to do with mains gas the authority is British Gas, acting through the appropriate region and local showroom. Advice at least is free, and they make it clear that they will make no charge if summoned to attend to a situation, usually a leak, where safety is involved. Any action subsequent to the first aid might of course become chargeable. Loosely allied to the gas industry there is for installers the organisation called CORGI, the Confederation of Registered Gas Installers. Membership of CORGI gives reasonable assurance that the installer has proved his competence to deal with all matters concerning gas.

Gas appliances, like oil burning ones, should be given regular skilled service. Though sooting is not an endemic problem, gas burners can develop faults of their own, like linting, which alter the carefully arranged gas/air ratio and affect combustion efficiency.

Gas appliances are in the main capable of working without a flue or chimney, and flueless room heaters and geysers were once quite common. In recent times however the practice has become restricted to appliances of low thermal capacity, and we do not recommend it. One of the bad features of an unflued appliance, gas as well as the crude portable paraffin heater, is the amount of water vapour it produces. Sometimes installed partly to combat condensation, it does in fact encourage it. In flue matters gas is the most versatile fuel, for as well as being able to manage without, it is equally at home with conventional, balanced, Se-duct, U-duct and assisted flues.

Though gas runs second to electricity in being easy to control in a great variety of ways, it is first in modulation. This is the type of control which can vary anywhere between on and off, smoothly.

Electricity

Electricity does not 'burn', is not a fossil fuel and so does not require a flue. But it is a fuel, since it is consumed and gives off energy as heat. In spite of the bitter warfare which occurs commercially between the fuels there is little going on in modern heating in which electricity is not present, even in a small capacity. Most control systems, for instance, are now electrical. A solid fuel boiler may be quite independent of it, but the circulating pump on the attached small bore system is not. Most oil fired boilers are wholly dependent. The point is worth making since, even nowadays, some people will weigh up electricity in terms of reliability — a throwback to the days when grid lines were often coming down. But it will be seen that failure of electricity will bring to a halt almost any central system, and perhaps the best way to insure against such a happening is to have at least one gas fire fitted.

Electricity is quite an expensive fuel. In compensation it is usually relatively cheap to install the necessary wiring; and it is very convenient, in being flueless, in appliances which are portable, and so on. There is a choice of tariffs for electricity users, and one, the off-peak tariff, owes its inception to electrical incursion into the central heating market. One other unique and disadvantageous feature of electricity is that it is unstorable. It must be made as required. But means can be found to turn it into heat first, and then store the heat. A suitable store must have several properties, which include mass and tolerance of high temperature. The first medium chosen was a concrete floor, from which came underfloor heating.

But the very mass of the subject rendered it difficult to control, almost impossible to do anything about a floor well warmed for a chilly day if Spring suddenly burst out — except open the windows wide. The system has therefore lapsed, to be replaced by the storage radiator and its big brother Electricaire. This type of heater in its variants is described at length in Chapter 9.

The ease of manipulation of electricity, the fact that anyone can run a cable and that the degree of expertise needed to run a gas pipe is not called for, has led to a good deal of controversy.

We have in mind particularly the type of system which is often claimed to be in some fashion Scandinavian, consists of several wall mounted heaters with controls, which will run for very low cost. If further identification is needed, it is sold by door-to-door salesmen, more often than not. And more often than not it is found to have two drawbacks. One is that the true running cost proves to be prohibitive, the other is that the instantaneous demand for electricity is beyond the capability of the average domestic service.

There may be genuine offers in this kind of system. But never, never commit yourself on the doorstep. Say that you will give a decision after you have shown the technical details to your local Electricity Board. No reputable seller can object to that.

Every electrical appliance has an opportunity to be tested for approval, and it follows that those most worth considering are the ones which have received approval. Look for the BEAB or Electricity Council label. When it comes to imported items, as increasingly it will, there is another consideration, that the item is suited to the characteristics of our current. It is not only having voltage in the range 230 to 250 volts a.c., but also having a frequency of 50 Hz (cycles), if there are moving parts.

There is no question but that electricity is not for taking risks with. Is it clear which connection is for earth, which for live? Does it have a fail-safe cut-off or similar device incorporated? If you are not absolutely sure about items of this kind, and did not buy the appliance from the Electricity Board (or even if you did and are still unsure) do go and ask their advice before making any sort of connection.

Electric heating of the heat storage type, which employs solid floors or storage radiators of the individual or central type, is designed for use with off-peak current. This is electricity generated in 'off-peak' hours, taken to be 11 pm to 7 am (or 2300 to 0700) with (in some areas) extensions beyond that period. It must be separately wired indoors, with its own meter, and unlike the standard supply does not have socket outlets. Each appliance designed for off-peak current is wired into the circuit.

Since the domestic hot water in a house with storage radiators is usually made by an immersion heater, and since this at standard tariff is more convenient than cheap, there would be considerable advantage in making hot water on the same off-peak tariff. This can be done, by arrangement with the Electricity Board. Since the basis of off-peak use is storage capacity, it follows that hot water made during the night must be sufficient for the day's use, and this almost always means a larger than normal cylinder. It is possible that arrangements can be made for a cylinder to be warmed by off-peak current, with a standby heater working at standard tariff in case some top-up heat is required during the day.

A much later tariff, the White Meter, is a very different type. It might almost be called a challenge. As though the electricity people were saying 'You've got a chance to win. But if you don't get thoroughly organised it is more likely that we shall win.' This is no bad thing, because getting organised is a sure step to economy — not only in gaining the advantage of the tariff, but in an all round saving of fuel. Getting organised means several things, the first of which, particularly for anyone with electric heating, means good insulation. It is greatly helped by planning for the automatic washing machine to perform in the small hours.

Cooking, for anyone with an electric cooker, is rarely other than an on-peak activity and weighs heavily on the wrong side of the balance. We have no figures which would show how the score stands, but we do emphasise that the White Meter tariff gives an incentive to do all the sensible things which perhaps would otherwise get put off.

Electric heating by low temperature radiant panels is not a form of storage heating and so does not qualify for the off-peak tariff.

Electricity has proved that it is capable of competing with the other fuels on their own grounds. But in addition, electricity has two special characteristics. One is the way other fuels are dependent upon it. The other is its self-contained, flueless role, which makes it specially convenient for the type of dwelling which is necessarily self-contained and with no outlet but its door. This describes the modern flat of the sort not served by an overall heating system, or a bed sitter.

Central heating by storage radiators is the only form of central heating of which it may be said, with practical truth, that you *can* take it with you. The value of the circuit, which you leave behind, is a very small part of the total.

Other Fuels

Two other fuels must be mentioned because they exist and are a practical proposition to someone, but in truth both are in a special and comparatively rare category.

There is wood, which used to be the only fuel available to our ancestors. Wood is nowadays in three classes. Anyone with access to woodland can get logs and sticks. Even when broken or sawn they are large units and best kept for the open fire. They are not suited to modern controlled apparatus such as a boiler, for several reasons.

(1) The density, i.e. weight of fuel per unit of volume, is low, and refuelling must be frequent.

(2) The absolute need to conform to a maximum size and often shape could impose an intolerable burden upon fuel preparation.

(3) Wood usually gives off during partial distillation a light tar, which coats heat exchange surfaces and lower flues.

Next to logs there are cases where someone has access to standard offcuts as a byproduct of some process. While this might relieve the second condition listed above, it leaves the other two unchanged.

Finally there is sawdust, a much more common industrial byproduct. This has a use as a boiler fuel, but not seriously on the domestic scale. Because it is tightly packed it needs a supply of forced air for combustion, and it is usual for factories producing sawdust to use it as fuel for their own boiler plant.

The short answer about wood is that it is not suitable for burning in standard solid fuel apparatus (other than the open fire) of the kind we are discussing in this book. But there are answers for those who are lucky enough to have supplies available at little or no cost.

Sawdust we have mentioned. But larger wood is still a standard fuel in parts of Scandinavia, and from there we import a limited amount of apparatus. There is at least one importer of Norwegian stoves, with a base in Wales and one in Scotland. And from Denmark at the time of writing there are two known to exist, one at least of them in Eire. The answer to wood must be, use suitable apparatus.

The second of the indigenous fuels is peat, widely associated with Ireland but still quite common in other areas, for example the 'mosses' around Manchester. Peat, like wood, is not suited to apparatus designed for coal and coke. It tends to be dense in the same way as sawdust is, but perhaps it may also be described as sluggish. There is a slow inevitability about peat which makes it easy to believe the tales of fires which have been burning for three or four hundred years. The Irish are taking a commercial interest in peat, but as a boiler fuel it must stay in the commercial and industrial region. It will not concern us here.

8 Automatic Controls

To anyone approaching the subject of central heating there are two kinds of controls. There are those you *must* have and so there is no decision to be made. And there is the other sort, which may again be put in two ranks, those which are to be strongly recommended, and those which if fitted will make some contribution.

The most anyone needs to know about the first category is how best to use them to obtain the most benefit, since like a willing horse they may need to be given guidance. The second category is more complicated since it brings in choice, decision making about whether to spend money, and so one needs to know what each type can contribute to comfort or economy or both.

Controls which are obligatory are built into the appliance. For example it is not possible to buy a gas fired boiler without a boiler thermostat and flame failure device; nor a pressure jet boiler without a control box and boiler thermostat. A further safeguard lies in the fact that if these parts were somehow to be removed from those boilers lighting would be impossible, because they are inextricably built into the lighting process.

A similar degree of interdependence exists in a free standing solid fuel boiler of comparable status. But it does not extend to the more elementary type of unit represented by the back boiler to either open or openable fire. This appliance always has a damper but it is manually controlled in most cases, offering no automatic safeguard against too much or too little fire. There are, however, models with thermostatic damper, and a model which has fan assisted draught is very readily controlled by thermostat. In the same type of appliance, where there exists a choice in the ratio of heat to air and heat to water, the controlling damper is manual.

Among obligatory controls the boiler thermostat is the only one which offers a chance to make routine adjustments. The easy solution is to ignore it, and leave it always on one setting, for example 70°C or 160°F. It can however be used with advantage, in the following ways.

(1) It must never be set below 60°C or 140°F.

(2) It should rarely be set above 90°C or 190°F, or in the case of solid fuel boilers 80°C or 180°F.

(3) Within those limits it may be used to spread the boiler's working time over the total 'on' time, to avoid long periods of being shut down on the thermostat. It is better for a boiler to work for 40 minutes in every hour than 20 minutes. To achieve this, adjust the temperature setting to the climate, for instance with maximum temperature in hard weather, tailing off to near minimum at each end of the heating season.

(4) If the system does not have a cylinder thermostat on the hot water cylinder, and there are very young or very old residents who might be at risk, avoid high temperature settings (in spite of (3) above) so that the domestic hot water stays at a hot but not scalding level.

(5) But note, in relation to (4), that if the dependents have the use of a thermostatic hot and cold water mixing valve, this works best when supplied with really hot water.

The cylinder thermostat is a device which senses the temperature of the stored water, and when it reaches the required figure shuts off the heat source. The thermostat which is incorporated with an electric immersion heater is a simple example. But the device is very often fitted separately, to control the water heating activities of a gas or oil fired boiler. This it does either by being wired back to the appliance so that it shuts it off when the water is up to temperature; or it operates a solenoid valve in the flow or return primary, thus preventing the flow of any more water into the cylinder.

The first method is neater, though on the face of it open to the objection that a satisfied hot water cylinder would shut down the central heating. By various ingenious ways not to be explained here, this is avoidable. The second method avoids it without prearrangement but introduces an extra solenoid valve, which would usually be fitted somewhere in the vicinity of the hot water cylinder. Referring back to Chapter 2 however, it will be clear that with solid fuel there must be no possibility of stopping flow in the primaries, since the cylinder, with or without towel rail, is the buffer or safety outlet against after-heat and over-run on the boiler thermostat control. It would be possible to arrange diversion through a three-way valve at such times, so that the primaries continued to flow through the towel rail only. But the cost and complication would rarely justify such a step.

A cylinder thermostat when fitted has a specific job, that of allowing the stored hot water temperature to be at its own best temperature, not dependent upon what seems best for the central heating. This means in broad figures that the hot water will stay at, say, $60°C$ or $140°F$ while the heating ranges up to, say, $80°C$ or $180°F$. It was a shock and an unresolved puzzle to become involved in an American offshore contract calling for exactly the reverse of that situation!

If the holding of a safe temperature of stored water is not important, it should be remembered that in times of heavy demand, as when a lot of guests want baths, that it is *thermal* storage that counts, not just gallonage. Thus, hot water at $80°C$ or $180°F$ will go about 25% further than the same water at $60°C$ or $140°F$, and if a cylinder thermostat is not fitted, this is a crisis which justifies using the boiler thermostat to boost the hot water storage.

The programmer serves both heater and heating system, and is a device of more or less operational complexity usually attached to the boiler, though it may be supplied detached, to fix to a convenient wall. The essentials of a programmer are a clock, boiler thermostat, wiring to pump and multiswitching. It is usual nowadays to use it as a junction for other items; for example the room thermostat which controls the pump and would otherwise be wired directly to it; cylinder thermo-stat; frost or low limit thermostat.

Usually the clock has four levers or tappets, which enable the user to choose two periods during 24 hours when the boiler shall start itself up and later shut down. The switching, reduced to a single knob, permits the user to decide whether, during the two working periods, he will have domestic hot water and central heating, or domestic hot water only. Note that unless the boiler has pumped primary flow (in which case the hot water system is effectively just like another radiator or heat emitter) the programmer will not permit heating without hot water. We have stated elsewhere our belief that most people manage very successfully with very few pro-grammes.

Elementary facts about using a programmer include such matters as noting the difference between night and day on the clock face, and turning the dial in the direction shown when altering the time setting. After a shut down for any reason, or a power cut, failure to reset the clock could lead to unexpected working hours.

The most commonly met control away from the boiler is the room thermostat. In most domestic premises there is one only, and it therefore has to act as the representative of the rest of the premises while being answerable only to the room in which it is fitted. This is not as chancy as it sounds, since the design will have allowed for the fact that while it may be set at say $15°C$ or $60°F$ in the hall, the emitters in the living room are of a capacity to bring that room up to $20°C$ or $70°F$, and conversely, supposing the roomstat to be in the living room and set to $70°$.

There are elementary common sense rules about the precise location of a roomstat. It must sense average conditions, not being tucked away in a corner where little air circulation takes place; not being chilled by being near a draughty door or window; not being artificially warmed by an adjacent lamp or television set or direct sunlight — or of course a radiator! Al-though the traffic through doors is a factor against using the hall, it nevertheless remains a very useful place, on the grounds that it is the one part of the house which should always be warmed. If the thermostat is in the living room this can never be allowed to go cold, without having an effect on the rest of the house.

The ordinary roomstat is an on/off device. But there is another, called a setback, which instead switches from normal to a low setting. It gives the user the ability to choose what the high and low settings are, and also the period of the setback, usually up to about 10 hours, after which it reverts to the high or normal setting. When there are special reasons for some warmth to be maintained all night, perhaps through illness, then the setback thermostat is well worth considering. The effect it produces with refined fuels, of a low level of maintained temperature, is almost precisely what occurs as a matter of course with solid fuel heaters which work at low level instead of going out.

Anyone who has a programmer has a clock, to control the times of operation. But for those who do not, a clock may be obtained separately, and for convenience wired into the room-stat circuit, since both are doing the same job, of controlling the boiler on/off. Other applications of a clock may be made, such as controlling the pump only if for any reason it is desired to keep the boiler always at work. Another example is that of zone control. This must start with the pipe runs, so that each zone is self-contained as a circuit off the main flow and return. A typical zone might be the first floor, the bedrooms, which have different needs from the living rooms. In that example, all heating would be cut off the first floor at, say 7 a.m. This would be the job of the clock, operating a solenoid valve. Such an arrangement is made into a unit in the Satchwell Minival.

Although the commonest clock is the four tappet type, variations abound. There is for instance day omission, which allows everything to stay dormant perhaps over a weekend, but brings it on automatically on Monday morning. Then there is the over-ride, a simple means of making a change of routine for one occasion only. Thus, an unexpected call out when an evening at home was planned, and the 'on' of heating can be over-ridden by a push button or similar. Next evening the heating will come on as usual without further action by the user since over-ride changes are self-cancelling.

Next on the list of instruments which are wired back to the boiler is the low limit thermostat, often called the frost stat. This is of great value to anyone in the habit of leaving the house untenanted for weekends or longer in mid-winter. One can of course keep the heating wholly at work, with thermostats set low. Over a week or two this is not the most economical of measures. The frost stat arranges that only when danger threatens will the system come on to work. The sensor is fitted at the most vulnerable place, where the first freezing might be expected, and the thermostat is usually set to a safe margin, such as $5°C$ or $40°F$. When this temperature is reached it cuts in, bypassing all other controls such as roomstat and clock, until the house is warmed sufficiently to satisfy it. A boiler which is fitted in an outhouse is an example of one which warrants a frost stat.

The radiator thermostat deserves serious consideration, because it makes up for the deficiency which we have seen in the single roomstat, that it controls all the house, except its own setting, by inference. If each radiator has a thermostat each room can be positively controlled. There are those who claim that this automatically leads to fuel economy, but there is an element of nonsense in that. It certainly gives the opportunity, but from that point on it is entirely up to the user. Since one result of intending to use a radiator thermostat is that radiators may safely be oversized (with a view to some temperature crisis) the user has the power to use more than normal heat. A pleasant feature of thermostatic radiator operation is that it is modulating. The water flow is progressively cut down to control the temperature, and this compares with the on/off operation in which a radiator alternates between hot and cold.

Radiator thermostats may have integral or remote sensing, but in either case it is necessary to keep this as far as possible from the influence of untypical temperature conditions.

The influence on the rest of the system must be taken into account. The use of radstats goes with a continuously running pump, stopped only by the clock during 'off' periods. One cannot after all have two controllers trying to do the same job. But there are installers who believe in putting a roomstat into the system as well, getting over the objection by setting it just above the highest temperature which would be expected from the system working normally on radstats. Presumably the roomstat then acts as a watchdog or longstop. Running experience of this method is hard to come by, and against it is the fact that radstats are not cheap, and the roomstat is an extra cost on top. Otherwise there seems to be no objection.

We must refer to the modulating or mixing valve which attracts a very mixed reception. The essential feature of this device is that it takes the boiler water, at boiler temperature, and allows it to pass to the heating system after mixing with a proportion of the cooler return water from the heating system. To put this another way, only a proportion of the water in the heating circuit is passed to the boiler for reheating at any given time.

In principle this is quite good, since if the proportions are right it is a means of adjusting the warmth input gradually to suit changes in ambient temperature. It is in effect a continuous operation of the kind we have suggested might take place seasonally, adapting the circulating water temperature to suit the prevailing weather.

It is plain that the mixing valve is not suited to being adjusted manually. That would be a full time job for someone. It must therefore be automatically controlled, by a thermostat. Where will the thermostat get its routine orders from? The logical answer is that since it is controlling the indoor temperature it will have its sensor indoors. But, say the objectors, this is too slow. The sensed change has already taken place, the corrective action will take time to work its way around and become effective. By that time it may not be needed and an opposite movement must be made, resulting in continuing oscillation of temperature, which is most certainly not the object of the system.

So those who favour the device now claim that the proper place for the sensor is outdoors, where its function is anticipatory. It feels a change in the outdoor temperature, and passes the information indoors so that the system picks up in anticipation of the effects being felt indoors. There is some initial difficulty in finding the best place for the sensor. Use a north wall, say the books, so that it is not foxed by direct sunlight. But suppose the house has a predominantly southern facing aspect, as it might do with a high pitched single pitch roof? Wind is as disturbing in its way as sunlight, so there must be some protection afforded − which the house cannot share though almost equally affected.

But the greatest objection, ignored by those who favour mixing valves, is a matter of timing. The indoor thermostat may be too slow in responding. But the whole of modern practice, as outlined in Chapter 10 (Insulation) is aimed at preventing outdoor conditions from reaching indoors − ever. What it succeeds in doing is to prolong the time taken and to modify the result when it does happen. Meantime the delay between the system receiving a signal and its effect becoming apparent is unchanged.

So it amounts to this: that a man may have a mixing valve at work which gives every satisfaction, bringing up the indoor heat release just in time to meet the oncoming cold spell. He then undertakes a big programme of insulation, which doubles the time of cold penetration and halves its intensity. And still the system reacts as before in response to the outdoor sensor. The result is bound to be at least as much overcompensating, and temperature oscillation, as the indoor sensor might give.

It might be possible to introduce a form of calibration into the system to compensate, though no one has offered to do this. The safest conclusion is that mixing valves seem a good idea but are perhaps best left out because of practical problems.

We do not include such items as automatic air vents in controls, since these are as fundamental as, say, the pump. Most of the useful forms of control are included in the above list, and if we may mention one other device which deserves to be omitted, it is the thermometer to measure air temperature. The only criterion of heating should be comfort, not a mathematical symbol unrelated to health, time of day or any personal factor.

9 Storage Heating

It is interesting to note that the principle employed in electric storage heating is likely to be applied to a much greater extent in the future. The principle is that of converting 'instant' energy, which if not captured when generated will be lost, to a form which can be stored, mainly as heat. Sources of instant energy include solar heat, wind, wave and tidal power and some more obscure sources. These will become more important as the supplies of natural gas and oil are depleted.

The reasons differ a little, though. Natural sources occur naturally, for instance the sun shines by day. But electricity is the same by day or night, except in cost of production. Hence it is cheaper to make it at night to use by day, for those applications in which this can be done; space heating is one of them. Electricity can be stored as chemical energy, which is the principle of the battery, but that is in another context.

The conversion of electricity takes place by 'burning' it, by hot wire as in an electric fire, in a situation in which all the heat generated passes into the storage medium. There is every good reason why the storage material shall be as compact as possible, if only because storage heaters already have a reputation for being on the bulky side. The two factors which govern the heat capacity are the weight and a heat retention factor called specific heat. Water, which has a very good specific heat, has been used, and iron billets, also lime and other substances for which some virtue has been claimed. But in the end practically all models come back to a ceramic material, which is either firebrick or silica brick. The latter has more of a tendency to spall or break with continual heating and cooling. Ceramics are not the best materials thermally but they are very robust and without noticeable expansion/contraction problems.

Thermal storage in its earliest form did not employ ceramics, but quite ordinary cement mixtures, used in the construction of solid floors. *Figure 9.1* shows the heating coil elements set between a solid concrete base below and a heavy screed above, with a layer of insulation defining the lower limit of heat storage. If less total storage capacity were required the insulation could be raised to the top of the concrete section if necessary.

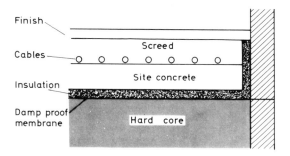

Fig. 9.1. Typical section through a heated floor

But of course such a decision could be made only once. Quite a lot of experience has been gained about the composition and application of screeds which resist the stresses of heating and cooling.

The more ingenious members of the public developed methods of operating which were very flexible (or so they claimed) but there is no doubt that underfloor heating has, in a changeable climate, the outstanding disadvantage of being inexorable. This combines badly with the anticipatory nature of the means of putting in the heat. If you decide that tomorrow will be cold you arrange for a greater charge to go in overnight. And if, as is by no means uncommon, tomorrow is not at all colder, the floor is still going to give off its stored heat, and there is nothing you can do about it, except leave all windows open or go out for the day. In converse conditions you are left to shiver.

This is not to decry underfloor heating for what it was — a pioneer effort which brought a lot of comfortable warmth to a nation shedding its spartan traditions.

The storage radiator is in many respects a notable advance upon the storage floor. It is not a constructional item, and so is available to anyone. It does not bring with it rules about avoiding the use of heavy carpet. Against it in small premises is that it is a piece of furniture, which the floor was not. But principally it is not, at least not necessarily, inexorable. There are in fact three classes of storage radiator. There is the common one, the cheapest, in which the heat leak is scientifically controlled by the amount of insulation given to it during construction, but is not otherwise controllable. This is comparable to the floor.

The other two classes of storage radiator have a common feature, that they are more heavily insulated so that random heat leakage is much reduced. Then, in the one type the heat output when required is damper controlled, usually manually. A damper is opened which allows air to pass from below through a passage or flue in the ceramic mass, where it becomes heated and passes out at the top. The second type of controlled unit is inactive until, under the influence of a room thermostat and probably a clock, a fan starts passing air through the heated core of the unit.

Either of the controlled types of unit is a great advance upon the first, in particular for quite common cases of a living room which is not much occupied until evening, by which time a natural unit is well past its best. The natural unit is quite well suited to halls, stores, libraries, and other situations in which bursts of warmth alternating with much cooler periods do not provide suitable conditions.

Another advantage to be claimed for controlled outlet models is that they relieve the user of a good deal of reliance upon a crystal ball. Whether tomorrow is going to be colder or warmer is less of a crisis if one may take the safe view and turn up the heat input, knowing that if it is not needed a lot of it will stay inside the unit until it is wanted, or it will form a substantial nucleus of heat so that the amount put in next night will be less. The maximum input is of course controlled automatically by a thermostat, adjustment of which is input control.

Inevitably the unit storage radiator grew up, into a central unit called Electricaire, big enough to provide warmth for several rooms. It is a large unit, and some are vertically disposed while others are horizontal. They are of the well insulated fan controlled type, the fan being powerful enough to propel the warm air through a duct system into the rooms to be warmed. In this connection it should be noted that standard units almost always have a stub duct system in mind, the fan being not powerful enough for a full duct system unless specifically stated.

Comparing an Electricaire unit with a stub duct unit using different fuel, the Electricaire has what might prove to be a decisive advantage in not needing a flue. As a heat unit it could suffer from its ability to become exhausted. That is, if given specially heavy duty during the day its heat charge could become used up before it is due to come on charge again. Instantaneous producers do not have this problem.

As a system it should be pointed out that it is unusual for an Electricaire unit to be connected to a return air duct, and return air is more usually allowed to find its own way back, if it wishes to do so. While this may work out quite well, it does tend to rely upon leaving certain doors open and is not therefore entirely automatic in its operation.

Electricaire units incorporate an air filter, usually on the inlet to the fan, and this should be kept clean. If it becomes choked, or even partly choked, it will affect performance and could be bad for the fan.

If we compare the warming action of a storage radiator with that of a hot water radiator, the nearest is the storage unit with uncontrolled output. Having less insulation it develops a generally higher surface temperature, and so is able to give off a higher proportion of invisible radiant heat. While this is an advantage it cannot outweigh the disadvantage of the uncontrollability of output; but when we speak of surface temperature we are speaking always of temperatures within safe limits. A part of the approval testing for any such unit is to ensure that its surface temperature cannot exceed a given standard figure. It should be noted, though, that some build-up can occur locally if normal heat dissipation is prevented, as it might be if say a pile of clothing were to be placed on the unit. Such units should always be kept clear.

We need to know what to do about the place where the unit stands. As a general rule nothing has to be done unless there are quite clear instructions given to the contrary. It is usual for all kinds of heating appliances to be tested for their effect upon the floor on which they stand, and in particular for the temperature which they will create below the unit. A storage radiator needs no special floor preparation. But while it is no doubt safe there is still the possibility that more or less continuous warmth will have some effect upon the floor covering. This may be most marked when the material is in the plastics or man made category, which includes floor tiles and many carpets. The way to avoid a tendency to softening of tiles, or embrittlement of fibres, is to use a protective sheet of asbestos millboard or similar under the unit.

The case of an Electricaire unit might be somewhat different, in that its gross weight might need support. We can only advise that in any given case the electricity people or the vendor must say what is needed, since larger units have greater weight. A suspended floor at ground floor level could for instance be given local support by means of brick piers taken off the under-floor: or the flooring could be cut out and a concrete base put in instead. In any case we recommend the use of an insulating plate under any such unit on a wooden floor, but stress that this is not for safety but to avoid any deterioration of the underfabric.

It is quite common for units of the Electricaire type to have an air duct between fan and heating elements at the bottom of the unit, thus forming a good insulator.

Boilers are a class of appliance which require some base protection, unless otherwise stated, and even then if there is a need to protect floor coverings. But the whole matter of boiler standing is covered by Building Regulations and to be on the safe side these should be consulted. The Regulations draw a distinction between solid fuel and oil fired appliances (Class 1) on the one hand, and gas fired appliances (Class 2) on the other.

The operation of storage units is, as already mentioned, anticipatory. Instead of using a crystal ball one may rely upon the weather forecasts, or if thought more reliable one's corns or piece of seaweed. But whatever means are used to predict the following day's temperature pattern, today is the day to do something about it. The input controller of a storage unit is usually marked in numbers, say 1 to 8. It takes a little while to discover, by trial and possibly error, whether a forecast of fairly cold justifies an input of number 5, or of 6. But this is what it amounts to, and any practical difficulties which may be met are rarely associated with that aspect. The input setting must be made before retiring for the night, since it controls what happens during the night.

The method is the same whether the unit is an individual one or a large Electricaire. The total reliance upon forecasting is greatly relaxed, as we have pointed out, by having units which do not rely upon natural heat discharge but to a very large extent upon extracting the heat required. The real responsibility then rests upon the owner to make sure that he programmes sufficient storage. To err on the generous side does not imply waste and loss since most of the surplus heat is retained by the insulation until it really is wanted. Prudence, economy and a proper regard for the conservation of energy require one not to be extravagant in the production of heat. But the logic of having a heating system in preference to putting up with the cold leads to the supposition that one will do what is needed for comfort. The guide line through this apparent dilemma is that which is set out in the Introduction, namely to have as much warmth as is needed, when and where needed; and outside those limits, nothing.

The service or maintenance on a storage unit may fall into more than one category, from ridiculously simple to the other extreme. Cleaning of the air filter on all fan controlled units should be done as regularly as inspection shows to be necessary; this will vary from unit to unit depending upon external conditions and the amount of dust which is in the air.

Beyond that an annual inspection may be thought desirable. Shut the main switch even though the current is off-peak and the time is on-peak. Remove sections of the casing and look in particular at the controller. Make sure that it is free from dust and that the points or contacts are not burnt. Check that electrical terminals are tightly screwed down.

In the extreme case that a heating element has broken, this can always be renewed in the case of a storage unit, whether single or Electricaire. In many cases it can be done without dismantling the core, though much of the casing must be removed. In the case of floor heating replacement is dependent upon construction. A floor may be made with a solidly embedded heating element, in which case nothing can be done. Or it may have a withdrawable element, with points at which withdrawal may take place. If taking over a house with this form of heating it is desirable to find out which type of element is fitted. Breakage of a heating element is an uncommon occurrence, but if it should occur with an embedded one there is little point in considering whether to dig up the floor. The easier course would be to change the type of heating, if only to storage units or Electricaire, retaining the off-peak meter and tariff.

10 Insulation

Whether you are more concerned to save money or running cost without loss of comfort, or to contribute to the urgent matter of conserving dwindling fuel supplies, the answer lies in insulation. It is not too much to say that it is more important than heating. If near-perfect insulation were possible, heating would not be required. Establishments such as the Centre for Alternative Technology at Machynlleth show how it is possible to save huge percentages of what might be considered normal household fuel bills, by a little planning in advance. This does of course include some planning in advance of building the house, a stage not available to everybody, but not to be neglected by anyone in that lucky position.

Insulation is a form of detective work, which to a large extent has been brought down to simple rules. Heat tries to escape, and the householder must find out where the leaks are, and stop them. We are by now fairly familiar with the tell tale, the house where snow does not settle on the roof in winter. The owner pays dearly to melt that amount of snow, and we all know that roof insulation would prevent it.

Roof insulation in the typical house or bungalow with pitched roof means insulation on the loft floor. It would take about twice as much, and a special rigid type, to insulate the pitched sides, though if the attic is brought into the occupied part of the house that is the way to do it. But if the attic is a place of dust and spiders then mineral wool or glass fibre, cut into widths to tuck cosily between the joists, is very suitable. It must not be laid under the cold water cistern, but may be carried up the sides and also tacked on to a wooden lid, as an extra precaution against the cistern freezing.

Vermiculite, a lightweight substance sold in granules, is another suitable material to be poured between the joists. After pouring it must be raked level.

143

The thickness of any such layer must be decided. When the subject was first popularised, those who advocated it were pleased if they could get people to accept a one inch thickness. Then it went on to become the 'economic thickness',

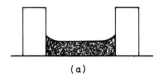

(a)

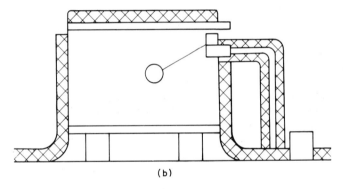

(b)

Fig. 10.1. Insulation in the roof space: blanket insulation (a) tucked between joists. The cold water cistern is contained within the insulated space (b). Wire netting holds the insulation in place

the amount you could put in and still recover the cost within reasonable time. It should be understood that the thicker the insulation the greater the saving, except that the benefits tail off a bit as the thickness increases. After all, if the first 50 mm (2 in) save say 60% it stands to reason that the second 50 mm cannot do the same. For a time economic thickness dictated the use of about 65 mm (2½ in) in the south, up to over 80 mm (3 in) in Scotland. That was before fuel prices soared. Now we may take it that a 100 mm (4 in) blanket is a good economic thickness, and it is of course a much better proposition than the original inch.

But insulation must be properly installed. This means being well tucked in, and in particular not allowing the wind, which always seems to blow in under eaves, to get under it or under any part of it and so carry away heat.

Suppose that you have to fit your insulating material on the pitch of the roof, it is possible to cut rigid or semi-rigid materials to fit between the joists as you would do for the floor. That is not however the best way to do it. Instead, take whole spreads of material and fasten them to and across the joists, so forming a complete surface of the material, as you would if fitting plaster board to a ceiling. By doing it this way you get the extra benefit of the air space between the material and the slates: or you may use that space to tuck in a non-rigid insulating material such as recommended for the loft floor. Or yet again you may use a cheaper front board, such as hardboard, and backfill the cavity formed with the insulating material.

After the roof come the walls, whose importance to heat loss lies mainly in their area. There are various ways of obtaining some benefit, even for instance planting a line of trees which shelter the wall from a strong prevailing wind, for wind encourages heat loss. A more uniform way of giving outside protection to a wall is by some form of cladding, it might be timber in the form of tiles or barge board, or slate or tile hanging. While these are not themselves outstanding insulators, they encourage the formation of a layer of still dry air, which is. In putting emphasis upon both 'still' and 'dry' we underline important principles. Referring to dryness, wall treatment will be less than satisfactory for any wall which suffers either from an inadequate damp course or from excessive porosity of the bricks. The damp course is rarely a problem with newer houses, and can be inserted, physically, or by chemical or electrostatic processes in older houses which do not have one. But brick porosity is relative, and is encouraged by driving rain and by north facing. It is only moderately expensive to apply a colourless preparation which will act to seal the front face of the brickwork.

Most houses nowadays are of cavity wall construction, having a cavity of approximately 50 mm (2 in) between the inner and outer leaves of brick or other building material. A ventilated cavity does not do much as an insulator, since it does not fulfil the condition regarding still air. An unventilated cavity is better from that respect, but does not always succeed in keeping the air still. If the cavity is wide enough there is room for an up-and-down circulation. The air in contact with the inner or warm wall will rise, causing the air in contact with the outer and cold wall to fall. The warmed air has to travel down the cold wall, where it loses its warmth, and in this way transfers indoor warmth to outdoors. Thus, good though air is at insulating, it is not as docile a subject as some solid material. This is where we arrive at cavity wall insulation, which in existing houses is injected through holes drilled in the wall and later plugged. The filling can take the form of a plastic material which after injection sets like a rubber sponge, being riddled with holes. It is the holes, which are not in contact with each other, which are still and dry and do the insulating, the plastic compound acting as a carrier for the holes.

The other type of material, used by Rentokil, is a mineral wool fibre which is blown into the cavity and packs it, doubtless with air entrainment. The plastic injection method can go wrong, for such reasons as having the wrong mix, and in spite of the great value of cavity insulation it is not cheap. Intending users would be well advised therefore to deal only with contractors who belong to the trade association, the National Cavity Insulation Association.

We cannot overlook walls in older houses which have no cavity but instead are perhaps 230 mm (9 in) or 330 mm (13 in) brick, but if we are to treat them by interior insulation it is even more important that they are not excessively porous. In the case of an inhabited house in normal use, with a single skin wall, the wall acts as a battlefield in winter. On one side is water trying to get in, and creeping through the pores. From the inside comes warmth to drive it out again; and somewhere about halfway, with luck, the battle reaches stalemate. The warmth has given up and become cold, but its efforts have halted the forward progress of the damp. That is why such a house, left empty for a few weeks, often shows damp on the walls, because no heat is being passed through from the inside. If we put insulation on the inside it will have the same effect, and damp can come right up to the insulation, unless it is checked at source. We have already described thermal inner linings when discussing warm air systems in Chapter 4 and there is nothing more to add.

The treatment of windows reached a ridiculous peak when it was taken up by people whose real interest does not lie within the heating industry. Ridiculous figures, like 50% saving, were freely bandied about, despite the fact that these would be obtainable only in a greenhouse, and then only with scientifically constructed double glazing. Anything with two panes of glass in it was and still is sold as double glazing for heat retention.

The ideal is still a sandwich of still dry air, which calls for a means of drying the air; and a sandwich width which offers a substantial depth of air but not enough to allow an internal circulation to set up, as we described happening in a cavity wall. The best distance apart for the two glass panes is in the neighbourhood of 18 mm (or from ½ to ¾ in). At that, the heat loss can be considered halved. The saving falls off somewhat as the gap decreases or increases, and by the time it reaches 100 mm (4 in) it is much more useful as a sound insulator than for heat. Since the price of proprietary double glazed units is so high many people are tempted to make their own, by adding an extra pane of glass. To be effective it must be done well, and the gap figure is given above. The handiest method of drying the trapped air is to introduce a little silica gel, a dehydrating agent, into the cavity. If this is properly sealed there will be only the original moisture to remove, not a continuing amount.

Questions are raised about triple glazing, which is not uncommon in parts of Scandinavia where the weather becomes really cold. But the diminishing usefulness of the extra pane or panes is very marked, and considering the current cost of even double glazing it could not achieve an 'economic' justification in this country. It should not be forgotten, though, that during the winter a good half of every 24 hours is darkness, and windows have no functional use. There is nothing to prevent us, then, from bringing in the equivalent of quadruple glazing, or more, for half the winter, by means of shutters, internal and external, by heavy curtains well tucked in, by any means we can devise to put a heat barrier across the gap in the brickwork.

The real importance of windows as a source of heat loss can often be exaggerated. In some houses they constitute a quite small proportion of the total wall area, and their greatest nuisance value comes from the condensation which forms on them. Double glazing, and of course the other steps mentioned, greatly reduce this. The greatest problem from windows comes to those who plan their homes with perpetual summer in mind, with great areas of glass as picture windows, french doors and the like. Almost prohibitive to double glaze, they can certainly do with ample heavy curtains. It is wise, when planning to incorporate such expensive items as large picture windows, to consider at that time incurring some additional expense and having double glazed units from the outset.

In the average house a quite surprising amount of heat is lost through the ground floor. In the case of a suspended floor this can happen in two ways: by conduction downwards and by the infiltration of cold air from below through gaps in floor boards. It should be remembered that a suspended floor must be kept well ventilated and no restriction placed on the access of air to it through the air bricks.

It will be apparent therefore that a solid floor offers more chances of success. It will not suffer from cold air infiltration, and during construction can have an insulating material built into it.

But for most people, who have to put up with what is provided, the first thing to do with a suspended floor is to stop up all the gaps through which air can travel. This may entail carpentry, or moderate gaps may be filled with papier mâché. Another sealer, possibly in addition, is to clad the whole floor in hardboard, wall to wall, sealing the joints with adhesive paper tape. Then in the case of suspended and solid floors, use floor coverings with a view to retaining heat; heavy quality underlay, and either good carpet or, if lino, then plenty of thick mats.

That deals with the top, sides and bottom of the box which effectively comprises any house, leaving the openings into that box, the doors and windows. To find a really well fitting frame to either door or window is exceptional, and again if the discrepancies are too great it may call for carpentry to put things right. This does not of course apply to metal fitments, which should fit unless warped by fixing stresses during erection. But sometimes metal doors and windows do not marry for the very simple reason that a blob of paint has destroyed the smoothness of the mating edge. It is worth going round each fitting to look for this. For moderate aberrations of fit in doors and windows the cure, in the form of some type of draught excluding strip, is fortunately not very expensive.

It was mentioned in Chapter 4, when discussing a fresh air intake to a warm air system, that the average house already lets in more than enough. It may well continue to do so even after treatment, but the probability has decreased. That is not a matter of vital importance, but the supply of air for combustion to any device in the house which burns a fossil fuel — coal, coke, oil, gas — whether it is the boiler or cooker, fire or warm air unit, is important. This gives us an opportunity to do things properly: instead of hoping that we live in a place so bad that in spite of ourselves it is self supporting in fresh air, we can now arrange to deliver fresh air to the exact place that it is needed, and in roughly the right amount.

There need never be serious misgiving about making a purpose built hole in a dwelling. Suppose that you put in an air brick or grille adjacent to the boiler. When the boiler is at work it will draw on it. When the boiler is not at work there will be no other force persuading air to enter through that aperture. If air is to enter, some must leave from somewhere, and you have already seen to it that there are no stray outlets. Indeed, the only stray air to enter a well treated house should be what comes in through the unavoidable opening of the front or back door.

It is quite easy to run a ducted air supply from a grille in the outside wall to a point adjacent to the appliance needing the air, but it is by no means certain that one should do so. A factor to be kept in mind is that in addition to heating there is ventilation to consider. In a sealed house we should gradually use up all the oxygen, and perish, and while there is little likelihood of this happening because of the near impossibility of achieving that degree of sealing, we must recognise it. In fact we do, numerically, since an allowance is made in heating calculations for a change of air in each room 1½ times an hour. Air movement within the house takes place of its own accord, by thermal currents and by the boosting caused by opening and shutting doors, and by the movement of people. The only critical aspect of this subject arises in providing minimum standards for combustion appliances, and figures for this are given in Chapter 6.

Continuing our search for sources of unofficial air leakage we come to the chimney which is not connected to a modern appliance with a sealed in flue connection. It may be that the chimney is not in use. In that case the least that should happen is that the register plate should be closed, and if no register is fitted then a blanking plate in lieu should be placed in the throat. The permanent way to deal with such chimneys is described in Chapter 6.

A fireplace which is still used occasionally should, if of the open fire type, first be reconsidered in case its importance warrants fitting a modern close-coupled appliance instead. If this is not done then quite clearly this chimney must be closed off in periods of non-use, either by a closeable register plate or by fitting a close screen across the opening. The amount of loss due to an uncontrolled chimney can be quite colossal. The effect of controlled leakage on the other hand is to give just that amount of room ventilation which is needed.

There is no doubt that a total programme of insulation, on the lines suggested here, will cost quite a lot. It is an unexciting subject, and unlike heating there is nothing to look at, nothing to burst into life at the turn of a switch, and a fairly common attitude is 'We'll have the heating this year, and do that next year — if we can afford it.' That is to put things in quite the wrong order, for if one had to make a clear choice insulation should win. We mentioned earlier that with ideal insulation heating would be redundant, and if we go only half way, we can say that with good insulation heating is less important. The other fallacy in the 'this year next year' argument is that heating plant needed *before* insulation is bigger and therefore more expensive than that which is needed *after* insulation. Further, if it becomes too large for the job its working efficiency goes down and creates another source of unnecessary running cost.

That argument was put into some very impressive figures recently, and it happened that experiments were carried out independently at about the same time by two quite uncon- nected organisations. Briefly what happened was that in each experiment two identical houses were built, but one was turned out in 'standard' form and the other was given a full insulation treatment. Each was then equipped with central heating, and since they were in the same area they experienced the same local climate. The two results showed remarkable accord, both indicating a saving of about 41% for the insulated house.

We would be surprised if any of our readers could emulate this, for this experiment was, as they say, under controlled conditions. But is there anyone who may at this moment be looking with disgust at his fuel bill who would reject the chance to knock even 25% off it? The matter does not end there. If the insulation is done first, the system will need a boiler or heater 25% (or 41%) smaller than it would have done, and the same will apply to radiators or other heat emitters. And cost goes along with size, so right at the start there is a chance to recover some of the cost of insulating.

Then there is an uncostable factor, the physiological one. Indoor warmth arrives in two forms, radiant and convected. Convected warmth is what is due to the circulation of warm air. Radiant warmth is a form of emanation which is not dependent upon being air borne. It may be visible, as from a flame or gas or electric fire, or it may be invisible if at low temperature. Equally important, it may be positive or negative in relation to the human body. That is, we may feel cold radiation as well as warmth, and both are more noticeable and more effective than convected warmth. Now, if we find ourselves in an uninsulated room with radiators, in which the temperature on a thermometer has just got up to the specified figure, we might still feel discomfort, and it is due to cold radiation off the still cold walls. Only the air is as yet up to temperature, and it will take quite a long time before the walls begin to absorb and retain warmth. But in a well insulated room, not only is the warming time for the walls less, but there is a much greater chance that the walls still retain a good deal of warmth from the last period of room occupation. That slight shiver in spite of the thermometer might never occur with insulation.

Just in case we ever get another prolonged spell of very hot weather, it is worth recalling that insulation also keeps out heat.

11 Domestic Hot Water

At various points throughout this book, where it was relevant to the matter in hand, we have mentioned domestic hot water. This is the hot water which is piped to bath and shower, to wash basins and to kitchen sink; very likely to washing machine and dishwasher too.

If the domestic hot water comes from an instantaneous gas fired water heater, or indeed from any existing plant such as a storage water heater which is in good condition, the rest of this chapter is unlikely to be of interest to the lucky owner. But for the rest we must bring under one heading what people should know about domestic hot water before choosing other equipment which might have a bearing upon it.

We could begin by asking how much water you need, and the answer lies of course with how many people use it, and for what purposes. It is less concerned with the size of the establishment, for we may find a castle holding only perhaps four people, while in a three-bedroom semi there might be seven or eight. Babies seem to need a lot of water. So do people who bath generously, while those who shower instead use far less. The main purpose in thinking about this is to make sure that you do not depart far from the average, and if you do you must do something about it. Do not neglect to include uses other than washing. A lot of people could mean a lot of laundry, and by hand or machine this uses water. A dishwater is also somewhat extravagant in its demand for hot water.

But having set that out, there is a figure which seems to suit the average premises, which we may take as being three-bedroom with what passes for an average family (2 parents and

2.3 children?). This is represented by a hot water cylinder of 170 litres (35 gal) capacity, with a heat supply of 3 kW (10 000 to 12 000 Btu/h). Obviously if we think about 4 and 5 bedroom properties with appropriate numbers in residence then we move to a 214 litre (45 gal) cylinder and up to 6 kW (20 000 Btu/h) make-up rate. For present purposes it is sufficient to think of 3 kW (10 000 Btu/h), and we would not recommend dropping below this level even if the premises is only a one-bedroom flat. With plenty of insulation on the cylinder the supply will last longer.

Ignoring some losses, 3 kW (10 000 Btu/h) will raise 45 litres (10 gal) of water through 55 deg C (100 deg F) in one hour. If we are starting with cold water, which is by no means to be taken for granted, then the recovery rate in the cylinder is 45 litres an hour. People may well ask why we do not aim for a better rate, to avoid so long a wait between baths. But first it would not just be a case of putting more heat in, for there are times when a central heating boiler of say 15 kW (50 000 Btu/h) rating is working only for the cylinder, and the rate does not improve. The limiting factor is the rate of heat exchange possible in the cylinder. Secondly, how much are you prepared to let hot water encroach upon central heating? Or in a slightly different way, how much extra would you be prepared to pay for a bigger boiler (and of course a cylinder with a special heat exchanger)? For in the first case, when both heating and hot water systems are at work, to take another 3 kW for hot water is to lose it from heating. We think we know the answer to those questions, which is why we now accept as general the 3 kW figure.

It is this which gives us that hardy standby, the electric immersion heater. With very few exceptions (of 2 kW or 2.5 kW) this is rated 3 kW (or it may be 3 kW/3 kW). The 3 kW/3kW is a very useful form of heater, suited only to cylinders which have the boss at the top, and it is made of two heater elements, one long and one short. These are separately wired to a change-over switch so that only one is at work at any one time. They are usually labelled Bath and Basin or something similar. When only small amounts of hot water are needed, as for washing up for instance, the short or Basin heater is switched on, and it heats only the upper part of the cylinder. For baths, or larger demands, the longer or Bath element is switched on and it heats water over most of the cylinder's height, as is usual with this type of heater.

Electric immersion heaters are the only form of heater which may be considered suitable for a direct cylinder, and indeed they would very often fail to enter fully a cylinder which contained an internal heat exchanger (*Figure 11.1*). Proprietary immersion heaters have a thermostat incorporated, and this is usually preset by the manufacturer. Some provide a knob for temperature adjustment.

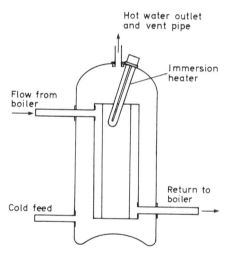

Fig. 11.1. Typical indirect cylinder into which an electric immersion heater has been fitted

There is little to say about the installation of this simple device, except that all wiring and switching must conform to IEE Regulations and these will be allied to local Building Regulations. But there is a feature of operation which crops up year after year, and with no hope of finality we deal with it now. The question is, is it more economical to leave an immersion heater on all the time, or to switch it on only when required? The answer is the latter. It may well be more *convenient* to leave it on all the time, but convenience is not economy. The reason is simple. Heat loss from one body to another proceeds at a rate proportional to the difference in temperature between them, in this case the cylinder and the surrounding air. Loss is also proportional to time: a cylinder which is always running is always (less a few operational dips) at maximum temperature and therefore at maximum heat loss. A cylinder is left to go cold loses no heat at that time, and proportionately less than maximum when cooling and warming.

The electric immersion heater owes its continuing existence to its great convenience. It has a low first cost, takes up next to no room, and is operated merely by switching on. Those factors count against the running cost, which is unquestionably high. But if we have a heating system of the electric storage type, which is charged at off-peak tariff rate, then it is advisable to ask that hot water be put on the same tariff, being another facet of storing. This will entail a change of method of operation, for off-peak current is available only during off-peak hours, say 11 − 7. All the hot water for next day must be made overnight, if the cheap tariff is to be enjoyed, and in most cases this calls for greater storage capacity, and a very determined effort to retain by insulation the heat which is received. These matters can be arranged by discussion with the local Electricity Board. It is not usual for the user to be left entirely at the mercy of his own operational calculation, though. The cylinder may have a second immersion heater fitted, connected to standard tariff current in the usual way, which may be switched on in case the hot water stock becomes depleted. Regular use of that device deserves to be looked upon as an admission of failure, and will certainly cut into any savings made on running costs after the expense of buying a larger cylinder.

This is an appropriate time to kill off another bogey concerning hot water cylinders. It is quite usual, and sensible, to build a cupboard around the cylinder, which has no visual attractions, and to make it into an airing cupboard. It is then not uncommon to assume that since it is now an airing cupboard some provision must be made to introduce warmth. This often results in part or all of the insulation around the cylinder being removed or omitted. There was never a more blatant case of taking a steam roller to crack a nut. Three simple facts must be appreciated.

(1) The cupboard is for airing, not for drying.

(2) With the best quality insulation it is likely to get, the cylinder will still lose enough heat to keep a mild atmosphere inside the cupboard.

(3) What most airing cupboards need most and lack most often is ventilation. This means, foremost, openings at top and bottom, so that warmed moist air can escape and be replaced by fresh air. It also means giving those items which are supposed to be 'airing' a chance to breathe, as they have when on the clothes line — not folded flat and bundled deep in a solid heap.

So do not skimp on the insulation for well intentioned but wrong reasons. This applies with equal force to any hot water cylinder, irrespective of the method by which it is heated.

The electric immersion heater with direct hot water cylinder is appropriate either to the special case of electric heating, preferably by off-peak storage, or to the very unspecialised area in which no other issue such as heating is involved. That is, the area in which a direct cylinder with immersion heater giving hot water is the only amenity.

When we come to consider the provision of hot water as a byproduct of a wet central heating system with boiler, taking our own advice and using an indirect type of cylinder, it is unlikely that an immersion heater could be used even if it were wanted. If it is desired as a standby it could be attempted, if the cylinder has a top entry boss. But you may ask why should it be wanted? Surely the boiler solves the problem of hot water quite satisfactorily?

It solves it, certainly, but not with entire satisfaction, as we shall show in a moment. But first to look at the physical details, which have already been touched upon in Chapters 2 and 3.

Hot water cylinders in this context may be indirect, or semi-indirect or self priming. If you already have a cylinder, the way to find out which of those two you have is to count the cold water cisterns which are connected to it. A fully indirect cylinder requires two cisterns, one for the primary circuit, which is continuous circulation of the same water, through boiler to cylinder and heating system and back to boiler; the other is the supply cistern for the domestic hot water, and in most houses cold water too (except at the kitchen sink).

A self-priming cylinder has only one cistern and one water supply, which sorts itself out inside the cylinder. It goes primarily to the secondary or hot water side, of course, but contributes make-up water to the primary circuit if this is wanted.

There is a structural difference which accounts for this difference in intakes. An indirect cylinder has full separation of the primary and secondary circuits, the former running through a coil or annulus which acts as a heat exchanger, the secondary water being on the outside. In a self-priming cylinder separation of primary and secondary is undertaken by a buffer of air, which is quite stable and adequate to keep the two apart over the whole range of normal operating temperatures and pressures. By the nature of the device however it must be recognised that a physical disturbance well outside the normal range could have the effect of breaking the seal, with temporary mixing of the two water supplies, primary and secondary. This would be very undesirable, since it would in effect return the system to all the drawbacks of a direct system. A disturbance which could bring about such a breakdown would be boiling, perhaps through thermostat failure on a boiler. Another possible cause could be a circulating pump which is far too powerful for the installation, creating great turbulence. It must be stressed however that such trouble-some factors are rare, and that self-priming cylinders correctly sized for the job they have to do are in very wide use and quite satisfactory. They are for open topped systems only, and cannot be used with sealed systems such as are described in Chapter 3.

For anyone who has a direct cylinder and wishes to change it there is a conversion set made, a coil type heat exchanger which is fitted into the cylinder by way of a boss intended for an immersion heater (which the coil is, though not electric). Such an arrangement is a regular part of a microbore system.

A fully indirect cylinder has four pipe connections: the flow and return from the boiler, the cold feed and the hot water offtake and vent pipe. In most cases the primaries connected to the boiler work by gravity. It is necessary therefore to see that their vertical progress is maintained. A pipe must never rise, then fall again before rising, for that way lie air locks which will soon shut the installation down. Even horizontal runs are to be avoided, partly because they can so easily lean a little the wrong way, partly because they absorb energy from the flow system. If some horizontal run is unavoidable it should be offset by at least three times the length of vertical run.

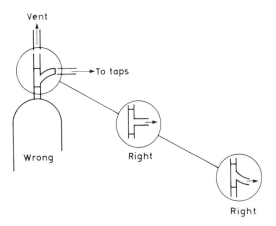

Fig. 11.2. The right and wrong ways of fitting the tee for hot water supply to the cylinder outlet

Another place to be on guard against an air lock is on the secondary flow pipe. The vent pipe must go upward as straight as possible, with no horizontal run at all. It must never be fitted with a valve. The domestic hot water is usually taken off this rising pipe, and it would seem quite natural to help this flow on its way by using a swept or pitcher tee, as shown in *Figure 11.2*. This is wrong, and will encourage air locking at that point. The correct fitting is either a straight tee or a reversed swept tee.

Now we will examine why domestic hot water as a byproduct of central heating is not the ideal system. We have already arrived at a figure of 3 kW (10 000 Btu/h) as the rate of heat input into the cylinder. We know that the present value of the rating of an average domestic boiler is in the range of 13 to 15 kW (45 to 50 000 Btu/h). This excludes the open fire with back boiler, referring to independent boilers mainly. Suppose we take a figure of 13 kW, then during the heating season the boiler is presumably well occupied, having to make 3 kW for hot water and 10 kW for space heating. Its work pattern on an average winter day may be 45 minutes on during every hour that it is at work.

The outstanding difference between the heating and hot water loads is that the latter is almost unchanged across the whole year. Whether the boiler is left for long periods in a potential on position, or whether it is switched on only for an hour or two each day, the fact remains that it is unable to deliver at a faster rate than 3 kW, or less than 25% of its rated output. It will therefore have long periods of inactivity in which to cool off, losing heat through dissipation and having then to reheat the structure of the boiler before coming into fully effective work.

The overall efficiency under these conditions is appalling. The reason why we put up with it is the common one, that it is convenient to do so. Although the logical answer is to separate the water heating function, and give it its own heater rated at 3 kW or thereabouts, this would (a) cost more initially; (b) take up more space, which is perhaps not available; (c) require another connection into the flue; so we do not do it.

Yet the objections are answerable, at least in the case of gas. The cost would be offset by the lower cost of a smaller heating boiler. A circulator, as a small gas boiler is called, may be wall mounted, and it may be obtained in balanced flue form.

A look at Chapter 5 will show that those air heaters which incorporate a means of making hot water already follow the recommended path, since the incorporated water heater is (with about one exception) operated quite independently of the heating, and is of the right size for the job.

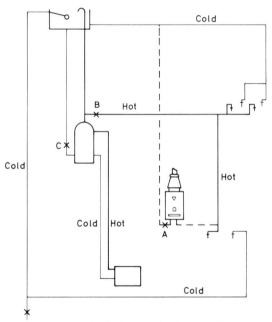

Fig. 11.3. The method of connecting in an auxiliary heater (for summer use, for example)

A variant of that method, and again the one which belongs to gas, is as before to keep the main boiler for space heating only: but to give water heating to an instantaneous heater, so doing away with the need for a hot water storage cylinder. It should perhaps be noted that this type of heater is forever heating and cooling off, but it is of very light weight construction and so not responsible for heavy losses as is a cast iron boiler.

Yet another method of dealing with the summer load combines both the others, and is shown in *Figure 11.3*. An instantaneous heater is piped into a boiler hot water circuit so as to be in parallel with the regular boiler. Heater and cylinder have valves fitted on one of the connections, so as to be able to stop any flow in one of the circuits. Thus valve A stops the inlet to the heater, valve B the outlet from the cylinder, and in each case the one valve is sufficient, and easiest to understand. This last is an important qualification, since there must always be reservations about handing out responsibility for valve operation to an unknown and generally inexpert public.

The method of connecting the instantaneous heater is preferably as shown by the broken lines, the water supply coming from the cold supply from the cistern. But there are bound to be cases in which the pressure, or vertical head, from such an arrangement is inadequate to work the heater, which has automatic valves and safeguards built in. In such a case the heater, being suitably constructed, may be connected direct to mains. But if that happens it cannot any longer be piped back into the existing hot water system. This would constitute making a cross connection between a high pressure and a low pressure supply, which is very properly contrary to the bylaws of the water authority. The existence of an isolating valve does not affect the matter. The requirement, that from the outlet of the heater a separate supply must be run to all outlet pipes, with their own taps, tends to make the thing so cumbersome as to be unacceptable. It would be better to go back to the scheme proposed, of removing hot water from the boiler altogether and giving it to an instantaneous heater.

The best working temperature for stored hot water, and the use of a cylinder thermostat to control that temperature, are matters dealt with in Chapter 8. Here we will repeat the warning that any means of totally isolating the primary circulation through the cylinder must not be used unless the boiler is capable of instantaneous automatic response to its own thermostat.

While so far we have shown how a gas fired system can be used to achieve better than average efficiency, we must come back to the fact that we live with the average, and that oil and solid fuel do not offer that degree of adaptability. Looking at solid fuel first, there are two levels of appliance. Numerically the most important is the fire back boiler, and there are undoubtedly many people who are forced to light a fire in midsummer because that is their only means of getting hot water. For them the standby immersion heater is the most convenient solution, though they would be able to use the *Figure 11.3* scheme if the conditions allow the heater to work.

The free standing and more controllable boiler is more adaptable. If the load is sufficient to keep it alight it may be left to slumber with occasional bursts of energy which keep the fire going. It is not in any better case than the fully automatic boiler in terms of working efficiency. It does not dissipate heat by intermittent cooling, but on the other hand it is always consuming fuel at some rate, however small. Although it would be difficult to obtain anything resembling an accurate costing, there is nothing here which justifies us in dismissing the electric immersion heater as a uniquely costly device to run. It does after all operate at 100% efficiency, near enough. A solid fuel boiler mainly slumbering, or a gas or oil boiler spending a lot of time shut off on its own thermostat, may well do no better than 50% efficiency. If we take a rough guess that electricity costs twice as much as any of those other fuels, the net cost comes out the same. The net cost ought perhaps to take account of extra wear and tear on the boiler, which is greater particularly under on/off conditions.

Although the purpose of this book is to consider hot water only from the heating point of view, leaving the plumbing for some other occasion, it is relevant to mention one aspect of hot water plumbing. This is the size of pipe to be used to convey hot water to taps. When the tap is open this pipe carries hot water; when the tap is shut the pipe is full of hot water, which slowly cools. When the tap is reopened the first stream of water is cold. This is called 'dead leg.'

All the rules of fluid flow indicate that we should use pipes of fairly generous proportions, so that friction is reduced and the tap is able to give a rapid discharge rate. But the more generous we are, the greater the volume of 'dead leg', with its accompanying inconvenience and sheer waste. Consequently hot water pipes are always calculated to be big enough but in no way generous in diameter. The desirability of planning to cut down the length of runs will also be obvious.

All the comments made about boilers apply to those mainly full scale back boilers built into combination units for gas or oil firing, and to the larger capacity hearth units which are the closed or closeable solid fuel room heaters with back boiler.

Instantaneous heaters are available for bottled gas, and are much in demand for mini-domestic conditions, in caravans and yachts. The only practical application for a domestic electric instantaneous water heater is in a shower, in which a very small quantity of water can be fashioned into a useful spray by a well designed rose. The limitation upon this type of heater is the electric loading which would be needed for say basin or bath filling − well beyond the capacity of a normal domestic meter.

Though we are considering domestic hot water as a service based upon a central heat source and a pipe network, this image does not have to be pursued regardless of common sense. It can often happen that a house or other type of dwelling is so constructed, or modified, that most of it lends itself to a compact pipe system: but that there is one part, perhaps a wing, an extension or the new cloakroom which used to be the laundry room, which stays obstinately outside any neat solution. If the system has to be extended to run hot water to a basin or shower it will mean a long pipe run, hot and cold, with nothing else on the way. So there is the cost of pipe and labour, the inconvenience, the continuing heat loss, the long dead leg, and particularly due to the last, a not very satisfactory service. This is a clear case for not extending the system, but for treating this outlet on its own, with saving of material, perpetual higher efficiency through less heat losses, and high user satisfaction.

It may be done in one of the following ways:

1.   By single point instantaneous gas water heater, possibly balanced flue type since the presence of a convenient flue would be a coincidence. This would involve running one cold water pipe, from main or from cistern, and a gas pipe; or in a non-gas situation having a bottle of gas nearby. This type of installation is adequate for a sink or wash basin, or for a shower so long as the shower equipment is carefully chosen to complement the heater.

2.   By electric storage heater of the smaller size, usually not more than 20 litres (5 gal) capacity. This may be of the free outlet type, to be situated above the outlet point; or it may be a pressurised heater which would stand under the floor, most usefully under the basin so that dead leg is almost nil.

Either of these heaters may be fed directly from mains or from the cold cistern supply. The free outlet or displacement type of heater is open topped and not itself under pressure. By opening a tap the user allows cold water to enter the heater, and this pushes out hot water. It must not have a control fitted to the outlet. The pressurised type of heater must first of all be bought for high pressure use. It must then be fitted either with an expansion pipe or with a pressure relief valve, and before undertaking to fit one of these the user should consult the local Electricity Board for detailed advice.

Up to the useful capacity, say 20 litres, either of these heaters may be used for a variety of purposes at any reasonable rate of flow. The free flow heater cannot be used for a shower since it must not have a mixing valve, a form of restriction, imposed upon its outlet.

The only services to be supplied are cold water and electricity, and perhaps in one case a vent pipe.

3.   The third possibility, useful for little but a shower or a basin spray tap, is the instantaneous electric heater which

we have discussed already. It would usually need to be 6 kW loading. Its requirements are a cold water supply and electricity.

Maintenance

The domestic hot water supply is inescapably at the mercy of whatever the local water supply has in store for it. We have shown how to protect the primary circuits by the use of an indirect system, excluding contact with continuous raw water. The domestic system has to suffer it, whatever it is. Things are not generally as bad as that may suggest. Troubles due to soft acid waters are rare since copper tube came into general use. The effect upon lead was of course potentially dangerous. But hard waters are different, since they do not attack their surroundings but simply deposit their hardness as solid matter when heated. If by the use of a chemical additive such as Calgon these deposits can be kept as a sludge to be washed away, not allowed to adhere as hard scale, then trouble with heat exchangers and pipes is long deferred. If not, the time must come when those items have to be defurred, and if this can be done by chemical descaler it saves dismantling the system. Signs of trouble are that the hot water gets less hot, and/or that the flow rate decreases.

The least harmful but most annoying symptom is probably the slight stain or encrustation which develops, like a junior stalactite, on bath or basin if the tap has a tendency to drip, perhaps when being closed. If the bath or basin is wiped down *after* dripping stops this may be kept off indefinitely. But if it does occur, it will usually be removed by being wiped with a cloth on which is a little proprietary scale solvent solution, then rinsed with clean water. But ask the bath manufacturer before applying this treatment, if you have a new and perhaps plastic bath.

To attempt to sum up this chapter:

1. Wet heating systems usually make provision in the size of heater to supply hot water as well as heating.

2. In terms of logic and efficiency this is not the best arrangement but is cheaper in first cost and takes less space than

3. The separation of hot water from heating by giving hot water its own heater.

4. A separate water heater cost is partly offset by needing a smaller heating boiler.

5. Means of providing hot water at about 3 kW rating include electric immersion heater, gas circulator; instantaneous gas water heater (higher ratings available).

6. Electric immersion heater goes with storage radiator system at off-peak rate.

7. Electric immersion heater may cost no more to run than a large boiler running at 3 kW load.

8. Hot water cylinder should always be well insulated.

9. The self-priming cylinder is a form of indirect cylinder.

10. Instantaneous multipoint heater may be fitted into a wet system for summer use on hot water service.

11. The use of a cylinder thermostat.

12. Avoidance of excessive 'dead leg'.

13. Domestic hot water for the remote part of the house.

14. Maintenance of the hot water side.

12 Facts and Figures

When all the decisions about type and style and so on have been made we come to the need for a calculation, to find out how big it will all be.

The calculation has two parts to it. First we have to decide how much heat is being lost from the building when it is kept, room by room, at the temperatures which are by now written into the heating specification, when the outdoor temperature is at a figure usually standardised at $-1°C$ or $30°F$. The amount of heat lost is a function of the building and has nothing to do with the form of heating used. It is therefore the same calculation for all occasions.

The second part of the calculation consists of putting the first part into practice. This means using it to choose the correct size of heat emitters in rooms; using the total heat loss figure to choose the right heat generator; and in the majority of cases, namely those in which there are physical links between generator and emitters, making sure that the interconnecting pipes or ducts are correctly sized.

The entire calculation process can be done by a rule-of-thumb means, but this is not recommended because it can make no allowance for variation of any sort. Another short cut which takes care of most of the variables is the mechanical calculator, of which the circular ones made by Mear and Co. are excellent. Because of their cost they are not usually included in the kit of a man doing only one calculation. So if we want to make a good job we have to do it the long way, which is by using first principles.

At this point let us make an appeal to amateurs and professionals alike. Keep it reasonable! Heat loss calculation is not a precise science, for too many variables are involved – a wind from the North, a feeling of tiredness perhaps in only one member of the household, and so on. It is therefore quite silly to go to great detail, coming up with a result (by no means uncommon in beginners) which shows an accuracy of perhaps 1 in 10 000. This is made even more absurd if then it has to be applied to choosing a piece of equipment which is not within 10% of the named figure, for lack of wide enough choice. So let us stick to moderate levels of accuracy, deciding beforehand what is, to use a mathematical term, significant.

To consider heat loss first: The Heat Loss Sheet shown in *Figure 12.1* is but one way of setting out the data and calculations, but it happens to be both clear and easy to use. One such sheet must be completed for each warmed room in the house in order to arrive at the total. The notes which go with it should be sufficient explanation of its use.

It will be seen that there are two elements in the calculation by this sheet. There is that which is unchangeable, and is mainly measured, being the dimensions of the room and its features – door, windows. In the same category, of being unchanging, are the two temperatures; the one you intend to have and the hypothetical outdoor 'low'.

The second element, which the sheet calls the U value, is the *thermal transmittance* and is a factor applied to the physical measurements, and it is changeable. The lower it is, the better for you. In metric units it is the number of watts (or joules per second) of energy which will pass through a transmitting surface such as a wall, over an area of a square metre, for each degree C of temperature difference between one side and the other. This is W/m^2 degC, or in imperial it is Btu/ft^2 h degF, where Btu is British Thermal Units and h is hours.

1 Btu/ft^2 h degF equals 5.7 W/m^2 degC.

In neither case is thickness mentioned, the U value referring to a standard area of the material *as it is*. If the thickness is changed, or some other significant difference made, a new U value applies.

HEAT LOSS SHEET

Design outdoor temperature −1°C

		Lounge			Dining		
1.	Room						
2.	Design room temperature						
3.	Design temp. diff. degC						
4.	Room length . .m						
5.	Room width . . .m						
6.	Outside walls, long. .m						
7.	Room height . . .m						
		Area	U value	kW	Area	U value	kW
8.	Outside walls, total						
9.	Windows						
10.	Outside doors						
11.	9 + 10						
12.	8 − 11						
13.	Unwarmed wall						
14.	Unwarmed floor						
15.	Unwarmed ceiling						
16.	Reqd. air changes						
17.	Room vol. 4 x 5 x 7						
18.	Air heat loss						
19.	Total heat loss						
20.	Allowance						
21.	Design heat loss						

If using Imperial units then: Design outdoor temperature is 30°F: physical measurements are in feet, and areas in ft². U values are Imperial: heat loss is Btu/h.

Fig. 12.1. The heat loss design sheet and how to complete it

To Use the Heat Loss sheet

Each room must be calculated independently.
The outside design temperature is usually taken as $30°F$ or $-1°C$
Then complete lines as follows.
1. Identify room, as kitchen, Bedroom 1 etc.
2. Design temperature for that room. See Chapter 1.
3. Difference between 2. and outside design temperature.
4.5. Longest wall and other wall.
6. Total of outside walls.
7. Height of room.
8. 6 x 7.
9. Area of windows, and U-value. kW column is area x U value x line 3.
10. Outside doors, as for windows line 9.
11. Add areas of 9 and 10.
12. Subtract line 11 from line 8 (area). Add U-value of walls. kW column, area x U-value x line 3.
13. If next door is a room or space at lower temperature heat will pass. Enter partition area, and U-value. kW column, area x U-value x line 3. If next door is unheated, assume halfway to cold.
14. 4 x 5 x floor U-value. kW column, area x U-value x line 3.
15. as 14 but for ceiling.
16. Air changes
17. 4 x 5 x 7.
18. Line 16 x 0.33 x line 3. (For Btu/h use 0.02, not 0.33.)
19. Add all the kW column figures.
20. An allowance may be up to 10% of line 19. We have indicated elsewhere our dislike of allowances, except in the case of selective heating, or where rapid warm-up is wanted and means exist to cut back to a 'normal' rate (as with thermostatic radiator valves).
21. Sum of lines 19 and 20 if applicable.

We must now refer to the chapter on insulation (Chapter 10) which was all about changing U values for the better. To take a good and important example, that of a 280 mm (11 in) cavity wall. Before insulation it would have a U value of about 2 in the metric scale (or 0.35 in the imperial scale). After cavity insulation this should be down to 0.6 (or just over 0.1), thus showing a saving of heat loss through outside walls of great magnitude. Similarly with double glazing, which when correctly done can reduce the window heat loss by half.

Table 12.1 SOME *U* VALUES

(In spite of the now speedy transfer of calculations to metric units we respect the preference of those who still think in Imperial. This table shows both sets of units)

    Note: only external walls have the low temperature on the outside. Internal or partition walls have some indoor temperature, and the heat loss is less though the *U* value is not altered by this fact. Total heat loss is a product of *U* x area x temp. difference.

Type of construction	U value	
	Metric	*Imperial*
External wall, plastered brick		
4½ in	3.2	0.57
9 in	2.4	0.43
11 in cavity unventilated	1.7	0.30
do ventilated	1.95	0.34
Window single glazed	5.7	1.0
double	2.9	0.5
Partition wall plastered brick		
4½ in	2.55	0.45
9 in	2.1	0.37
Pitched roof		
Tile on board & felt	2.0	0.35
Plasterboard ceiling, roof space		
over: 2 in vermiculite infill	0.8	0.14
Ground floor		
Suspended, single air brick	1.7	0.30
ditto with parquet, lino etc.	1.4	0.25
Solid floor on earth	1.15	0.20

The next question is, how do we find U values? These cannot be measured in the way wall areas are measured. But a great deal of experimental work has been done, resulting in tables of values which are true within the accuracy of the subject and cover all standard building materials. The best source for a table is in the *Guide to Good Practice* published by the former Institution of Heating and Ventilating Engineers, now the Chartered Institution of Building Services. The Guide should be obtainable through a public library. *Table 12.1* lists a few U values.

In one area U values are not tabulated. This is where new and specialised applications of insulation are concerned. Again a good example is that of cavity walls. Since the insulation applied is of greater importance thermally than the brickwork (which is 'standard'), it is the insulation which determines the U value. For this, then, we have to rely upon the maker, who will supply a figure. But in broad terms the result will be in the area already mentioned.

Perhaps the make-up of the calculation sheet will emphasise the way in which U values are integral in deciding upon sizes of apparatus. In this way it will add extra weight to a point we have made already, that insulation work is not something for next year, but the very first part of a heating installation. Only in that way will it be possible to gain the greatest advantage in offsetting the cost of insulation, and to start from scratch with an installation which runs at minimum cost for maximum results.

When heat loss sheets have been compiled for all rooms to be warmed, we have a complete picture of the house's heating requirements. We can use the figures first to decide about heat emitters, room by room. Let us suppose that we have to deal with a room whose heat loss is calculated as 2.452 kW or 8386 Btu/h. The first thing to do is to turn these into sensible figures, 2.5 kW or 8400 Btu/h. (You will note that we always round *up*. One result of this is that given alternative apparatus, one being just below the datum figure, we can often accept it in preference to one which is well above the line.)

Now to fit our figure to some types of heat emitter.

Hot Water Radiators

We will begin by assuming that consideration will only be given
to makes which have MARC approved outputs. If you feel free
to look at all the makes covered by this, you will no doubt find
one radiator which is very close to the heat loss figure. But
you may be restricted, by availability or by personal preference
for appearance. We have restricted ourselves to one maker
only, in quoting the following:

(1) a single panel radiator 685 mm (27 in) high x 2.77 m
(109 in) long, with output 2.5 kW or 8545 Btu/h.
(2) a double panel radiator of the same length but only
381 mm (15 in) high, output 2.47 kW or 8430 Btu/h.

Either would be quite suitable, supposing that you have a wall
length of 2.77 m available. If you do not, then apart from
looking at other makes you could split the duty between two
radiators in the one room, which is well worth considering if it
is a long or L-shaped room. In such a case two single panels
each 685 mm (27 in) high x 1.35 m (53 in) long would give
2.46 kW or 8390 Btu/h between them. Or of course you could
use two panels of dissimilar output, adding up to the required
total.

Note that if the system is to have thermostatic radiator valves
fitted, larger panels may be used and the size is less critical.

Skirting Heating

This is bought, rather like string, by the metre or yard. Any
given make or model has a stated output per unit of length,
and to obtain a known output is a matter of ordering enough
length — and of course having enough skirting length to fit it
to. This can mean that when measurements become critical
you will have to shop around for the best output per unit length.

In our example, and taking a typical heater of 450 W/m run,
the length needed would be 2.5 divided by 0.45, equal to 5.5 m
run.

Natural Convectors

These will be chosen by much the same means as apply to hot water radiators, with the exception that the choice is very much less, so that oversizing with reliance upon controls to reduce the running output is much more common.

Fan Convectors

Fan convectors are almost wholly thermostatically controlled. Oversizing is within reason a good way of achieving rapid warming of cold rooms, without introducing other problems with the heat generator. Unless a fan convector has fan speed control, however, too much oversizing can lead to long periods of automatic shut-down, which is not a desirable condition. To refer it to our example by way of typical marketed units, we find one which gives, from Boost fan setting down to Off outputs of 3.3, 2.3, and 0.46 kW (11 400, 7 800, 1 600 Btu/h): another, with more fan switch positions, gives 2.6, 2.3, 2.04, 0.35 kW (9 020, 7 980, 7 060, 1 200 Btu/h). Either would be well suited to the duty.

Unit Heaters

Gas fired unit heaters are independent units, connected only to a gas supply. The choice of size is next to nothing, except from make to make. Usually the most one can hope to do is to obtain a unit which is not too small for the duty, relying upon automatic controls to take care of the excess of output whatever it may be.

Electric unit heaters are more usually known as storage radiators. First a word of warning. Do not be misled by the standard terms in which these are named. Thus, a 3 kW electric fire has an output of 3 kW. But a 3 kW storage radiator has an *input* of 3 kW. We know only that it will take in enough energy for a fairish stint — more than a 1.5 kW model will, for instance.

In order to find out what it is capable of giving out, we need to know the total energy input. Over an 8 hour charging period this would be at maximum 8 x 3 = 24 kWh. Supposing that the rate of discharge were even it would give, over 24 hours, 24 into 24 or 1 kWh hourly rate. But as we know, the output is higher than the average when freshly charged, and lower towards the end.

Controlled output storage units

These should not need to differ in rating, i.e. in input. Their characteristic is that of a different time distribution; not a descending almost straight line graph but a low graph line with peaks where the fan or damper operates. The extra heat given out in On periods is that which is conserved during the Off periods.

Having made those points about sizing storage radiators however we recommend that a final word of advice should come from either a chosen manufacturer or from the local Electricity Board showroom, since units of this kind may incorporate features which affect capacity, retentivity or some other aspect.

Warm Air Registers

Though the recommended air exit velocity is not the same for all positions of the register (in floor, wall, ceiling for example) it is broadly in the range 1.0 to 2.0 m/s or 200 to 400 ft/min. The register size is then related to this velocity: to the volume of air to be handled, which is in effect the amount of heat being handled: to the free area of the register (usually about 70%). The net result of combining these factors, as manufacturers do in tabular form, is to enable a correct size of register to be chosen without trouble. Return air grilles, though not fully comparable, will be roughly the same size, having to handle the same air volume.

Heat Generator

Just as there are variants of the wet system so there is more than one correct assessment of the size of generator, in this case the boiler. We will assume that the system supplies domestic hot water as well, for which an allowance of 3 kW plus or minus 20% depending upon estimated demand is made (10 000 Btu/h). It was noted earlier that we do not favour the addition of a 'margin' or allowance for contingencies except in the case of non-hopper type solid fuel appliances. In their case the contingency, of fluctuating firebed, is a permanent feature.

Full central heating systems, partial and background systems, all subject to simultaneous demand, should have their boiler sized to take care of the total calculated heat loss, including the hot water allowance.

Selective systems justify their existence by having a smaller, therefore cheaper boiler. It should be able to deal with the load from the greatest number of appliances which it is intended will be at work at any one time. Whether this total should include the hot water allowance must depend upon a decision whether the making of hot water should be a front line activity or whether it should occur mainly in lull periods. There is a further capital economy in taking the second view, in which case no allowance would be made. Another case in which no hot water allowance is made is that in which priority switching controls are included in the system.

Heat generators for microbore systems are chosen on the same lines as for small bore systems, except that it is more usual in microbore systems to arrange the hot water through a pumped coil in the cylinder, when it ceases to become a different type of circulation and instead is effectively just another radiator. In calculating boiler size it should be taken into account among the heat emitters.

For warm air systems the domestic hot water, if applicable, is made in a separate piece of apparatus and does not enter into heater size. The rules for choosing a heater are as stated for wet systems in showing a difference between selective heating and other forms. We would urge the view that great reliance should be placed upon selective heating in this context. The process has already proved that it is economical to install and to use, and in addition it is easy to control room by room, shutting and opening registers. It should therefore have every opportunity to benefit from such other economies as are readily available.

Unit heaters are their own generators.

Electricaire is a central unit dealing with heating only. Using the type of calculation indicated for storage radiators it is possible to find a suitably sized unit for the duty, selective or otherwise. But this is again a case in which we recommend seeking advice from the manufacturer before plunging.

Transmitting Heat

The first step is to decide where the pipes or ducts are going to run. We have already dealt with the technical aspects of perimeter circuits, back-to-back appliances, one and two pipe circuits, the 'out-and-home' runs of microbore, and such matters as zoning. The conclusions reached should now be put on a plan of the house, floor by floor, and studied to make sure that the runs chosen are as direct as possible, with a minimum of bends, rises and falls, long spurs and other factors which would either add to resistance to flow or cause some appliances to be isolated at a distance from the main or index circuit. In the case of warm air it is not a circuit, but the principal feed duct is called the index.

Once a satisfactory drawing has been made it may be annotated with quantities. Starting from the emitters we can show what each pipe must carry in heat units, and where two pipes branch the feeder will acquire both quantities. Ultimately the index circuit or duct will be responsible for the total, though not necessarily in all cases for its whole length. This will be clear from your diagram.

We have made our own diagram (*Figure 12.2*). It is very elementary, omitting valves and all other detail not related to the purpose in hand, which is to explore the basic circuit. As you will see, we include two zones, C and D, and omit domestic hot water because it is not part of the circuit. The circuit shown

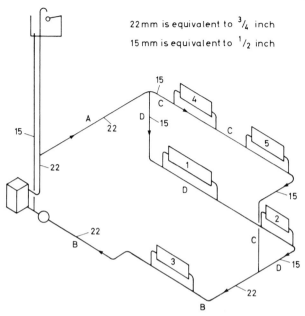

22 mm is equivalent to ³/₄ inch

15 mm is equivalent to ¹/₂ inch

Fig. 12.2. Basic layout of radiator circuits

Table 12.2

1.	2.3 kW	7 900 Btu/h
2.	1.6	5 000
3.	1.7	5 800
4.	1.4	4 800
5.	1.8	6 000

Table 12.3

Circuit			
A:	0.21 kg/s	1 500 lb/h
B:	0.21	1 500
C:	0.077	540
D:	0.094	670

Table 12.4 METRIC COPPER PIPE

Frictional loss per metre run for a range of flow rates (kg/s) and pipe diameters (Part of a larger table, reproduced by courtesy of the Copper Development Association)

Flow kg/s	6mm N/m²	8mm N/m²	10mm N/m²	12mm N/m²	15mm N/m²	18mm N/m²	22mm N/m²	28mm N/m²	35mm N/m²	m/s
0.074			2130	810	273	113	41.7	12.0		
0.076			2230	848	286	118	43.7	12.6		
0.078			2330	890	298	124	45.6	13.2		
0.080			2430	930	312	130	47.7	13.8	4.85	
0.084			2640	1010	341	141	52.0	15.0	5.3	0.1
0.088			2850	1100	368	153	56.3	16.3	5.75	
0.092			3090	1180	397	165	60.8	17.6	6.23	
0.096			3330	1270	430	178	65.5	19.0	6.70	
0.100			3570	1370	462	192	70.5	20.4	7.20	
0.105			3890	1490	502	208	77.0	22.3	7.55	
0.110			4200	1620	545	226	83.5	24.2	8.55	
0.115			4550	1740	588	245	90.5	26.1	9.25	
0.120			4900	1880	633	264	97.0	28.2	9.95	
0.125				2020	680	283	104	30.2	10.7	
0.130				2160	728	303	113	32.4	11.5	
0.135				2310	775	324	121	34.6	12.3	
0.140				2450	828	345	128	36.8	13.1	
0.145				2600	880	366	136	39.3	13.9	

							0.2
0.160	3090	1040	434	162	46.6	16.4	
0.165	3260	1090	458	171	49.3	17.4	
0.170	3430	1150	483	180	52.0	18.3	
0.175	3600	1210	508	189	54.8	19.3	
0.180	3780	1270	533	199	57.5	20.3	
0.185		1340	560	209	60.3	21.3	
0.190		1400	587	218	63.0	22.3	
0.195		1460	613	229	66.0	23.4	
0.20		1530	642	240	69.0	24.4	
0.21		1670	700	261	75.5	26.5	
0.22		1810	755	283	82.0	28.8	
0.23		1950	820	305	88.5	31.2	
0.24		2100	880	330	95.5	33.7	
0.25		2250	950	355	103	36.2	
0.26		2410	1020	380	110	38.7	
0.27		2580	1080	405	118	41.3	
0.28		2740	1150	432	126	44.1	
0.29			1230	457	134	47.0	
0.30			1310	485	142	49.7	
0.31			1380	513	150	52.8	
0.32			1460	543	159	55.8	0.4
0.33			1540	574	168	59.0	
0.34			1630	606	177	62.0	
0.35			1720	637	186	65.0	
0.36			1800	668	195	68.5	

is the single pipe type. We are taking the ratings of radiators to be as shown in *Table 12.2*.

Since in this case the lengths of circuits C and D are very similar, the index circuit, i.e. the one offering more resistance, will be ADB, where D includes radiators 1 and 2, and radiator 3 is in B.

We can now find out, first how much water each leg must carry, and from that the required pipe size. In British practice it is usual to allow for a temperature drop across the circuit of 11 degC or 20 degF. Since the figure is not critical we are taking 10 degC. The amount of water passing comes from

$$\frac{\text{Heat required}}{\text{Temp. drop x specific heat of water}}$$

The heat required is in watts or Btu/h: the temperature drop in degC or deg F: the specific heat of water is (metric) 4180 or (imperial) 1. Using that expression with the figures in *Table 12.2* we arrive at *Table 12.3*.

Table 12.5 IMPERIAL COPPER PIPE

lb water/h ½in	¾in	1in	Friction: in w.g. per ft run
200	580	–	0.06
250	720	1620	0.09
300	800	1860	0.13
350	1080	2280	0.16
400	1200	2540	0.20
450	1320	2760	0.24
500	1380	3000	0.28
550	1560	3240	0.32
600	1650	3480	0.40
650	1920	4080	0.48
720	2020	4320	0.54
780	2160	4620	0.60
840	2400	5100	0.72
900	2580	5580	0.84
960	2820	6000	0.96
1020	3000	6480	1.08

With the aid of a table showing the resistance to flow of copper pipe at various loads, *Table 12.4* or *Table 12.5,* we can go on to find the correct pipe sizes. These must be capable of passing the required amount with a velocity not exceeding 1 m/s or 3ft/s. These tables differ slightly in construction, but in both cases the principle is that the selected value must lie *above* the solid line which is stepped through the table. In *Table 12.4* there are several such lines. Considering only the index circuit, and reading *Table 12.3* along with either *Table 12.4* or *12.5,* we can now make up *Table 12.6*, below.

Table 12.6

A.	0.21 kg/s	22 mm pipe	261 N/m²/m	8.3 m	2170 N/m²
B.	0.21	22	261	9.9	2590
D.	0.094	15	420	17.7	7450

This shows us not only the correct pipe sizes but, in the last column, the total resistance to flow which the chosen pump must be able to handle. The equivalent table for imperial units is

A.	1 500 Btu/h	¾ in	0.32 in w.g./ft	25 ft	8 in w.g.
B.	1500	¾	0.32	30	9.6
D.	670	½	0.54	54	29.0

(The lengths, of 8.3 m etc. and 25 ft etc., have been introduced in order to show the resistance calculation.)

It is safe to say that all small bore practice operates in the 15 to 22 mm (½ to ¾ in) range, with occasional shifts to 28 mm or 1 inch in larger establishments. It is quite common to make the primaries to the hot water cylinder in 28 mm or 1 inch particularly if the runs are at all difficult, since this is not pump pressurised.

The same principles are used when calculating for a microbore system. The practice to which it refers shows some slight difference. In place of one or two circuits there is a number, and these must be evaluated for resistance to decide which is the index. Then, the pipes are much more in the 6 to 12 mm range, mainly 8 and 10 mm, and rarely coming up to 15 mm, except of course for the manifolds. Resistances are understandably higher, and this accounts for the use of a pump in a higher pressure range than for small bore.

Warm air systems: the principles already outlined apply, in cases where the vehicle for heat is air instead of water. We design a duct system after finding out how much air must be carried in each section of the ductwork, then relying heavily upon makers' tables to choose the duct sections which will carry that air with due regard for velocity and pressure loss. Ducts differ favourably compared to pipe in offering a comparatively large range of sectional areas. That is why it is possible, as stated earlier, to have a main duct stepped down as its length proceeds, in order to maintain the velocity despite diminishing volume.

In consideration of the somewhat esoteric nature of a full duct system, that it applies almost exclusively to new building, we will not go into details about the calculation. A stub duct system is rarely the subject of a serious calculation. If it is made for its short length to the section of the outlet port on the heater it achieves its purpose and the responsibility for regulation of air flow is given to the outlet register or registers.

13 Choosing a System

You have decided (let us say) that you will go in for central heating; or that you will get rid of the old and worn out system you have, for a new one. On grounds of cost alone it is not a step to be taken lightly. But it is just at this point too that all the dreadful stories come to mind, of people beggared and bankrupted, made to wait two years for essential parts, driven to desperation — and all they started with was a simple wish to keep warm.

Well, no doubt about it, you must keep your wits about you, or else employ someone who can be fully trusted to do the caring on your behalf. For a start you can count those who will take a negative attitude to your job:

The local authority, who will want to increase your rates.
The local authority, who will want to see that their Building Regulations are not breached, by way of structure, or fire risks, or electrical details.
The water authority, watchful of their byelaws.
The Clean Air people if you burn coal or oil.
In certain circumstances you might even come up against Planning, e.g. if you want to put up an outhouse for a boiler.

But all on that list, and any we may have missed out, are really quite helpful if asked for help. What tends to upset them is having their rules broken without prior consultation which would have avoided it. So the first practical rule should be, to make a list of everyone who has a genuine need to know what you are doing, and to tell them.

No doubt you will then ask how do you know who these various authorities are? So we go one step back, to the chosen installer or to the fuel supplier who may be advising you about the new installation. He spends his working life in this area of activity and should know. He should know also any short cuts. This does not mean evasions. But suppose for instance that a national agreement has been made that a certain form of appliance detail is acceptable in spite of earlier precedent, then this need not be argued again. That kind of detail is for the fuel supplier or the appliance supplier to know.

The second practical rule is, therefore, to discuss the proposals in some detail with a representative of your local fuel supplier, Gas Region or Electricity Board, major oil company (Shell, Esso etc.) agent, or SFAS or NCB agent. The chosen one of these stands to make a continuing profit out of you, and this is your turn to get something besides fuel from him.

But, you may well say, we haven't yet decided which of these we will use. True. So go back yet another step, and see what is going to influence the decision between the kinds of heating which are outlined in this book. Here are just a few points which might be applicable.

(1) Age. Moving into retirement, do you seek simplicity and ease? Will you want to carry coal and ashes? Will you want to lie on the floor to light a blown out gas bypass flame? Or will you insist upon simple controls at reachable height all the time, and no humping?

(2) Physical disability creates the same conditions as age, above.

(3) Your premises, are they rented, owned, do you plan a long occupation or intend to sell and move in say five years? For a start you would not plough a lot of capital into a landlord's property unless there was a beneficial arrangement with the landlord. No need to go without heating, though. This is a situation suiting storage radiators, of which you can remove all but the circuit; or the hearth mounted oil fired unit which gives bulk warm air distributed around the house

by drift and perhaps an extract fan. It is *not* an occasion for running a pipe circuit or extensive ducts.

There may well be a different outlook towards the place in which you hope to spend the rest of your days, compared with one which you expect to move on from, as is common enough in these restless times. The case is however debatable. On the one hand it may be thought that to put in the best system available will bring its own reward in the price which the house will fetch eventually. On the other hand, pessimism or some other reason may dictate that you put in the cheapest form of system compatible with being able to describe it as central heating, and to enjoy it during occupation of the house.

(4) Geographical location. Fuel limitations sometimes apply. In quite a few places, usually smallish and quite a way from anywhere, there is no gas main. Such places almost always have a Calor gas agent. Only rarely do we find a place with no electricity — a cottage on the moor, perhaps. This is extremely limiting, because of the great reliance of so many other appliances upon electricity for their instrumentation and control. Possibilities coming readily to mind are a solid fuel boiler (but no circulating pump for a wet system); an oil fired hearth fitted heater as described in (3), in which the extract fan is not an essential part and is replaced by plenty of open doors. Pity the cottage-on-the-moor dweller if he is inaccessible to a supplier of solid or oil fuel. He has little left but wood, or peat, or solar heat, none of which come into this book. The portable paraffin stove is not an exciting prospect, since it gives off a gallon of water for every gallon of paraffin burned, and is a chief contributor to internal condensation.

(5) Household detail. This covers a good deal. It covers the flat which has no flue and no right of access to an upper floor, for which there is little choice beyond electric storage. It covers the new house built for the electric age, with no chimney, which does not prevent you from using gas fired balanced flue appliances; or from building a flue for yourself (subject to Building Regulations).

It covers the place which has no spare floor space, e.g. a 'bijou' kitchen, and the under stairs cannot possibly be connected to a flue. Electricity apart, there is the wall mounted gas boiler, and skirting heating to save the space taken by radiators. Or, given a hearth, there is the back boiler. Still under household detail we must think of places which have solid floors, and ask how will we run the pipes of a wet system. Perhaps tied to the same set of conditions will be a strong wish to keep all the 'engine room' details out of sight. The primaries of the hot water system can usually be accommodated in the core of the house, particularly if the hot water cylinder is vertically over the boiler position, as for good operation it should be. We have also drawn attention to the latest idea, in which the hot water cylinder is a part of the boiler and so primaries are done away with. But the heating system is different. It must run outwards to the heat emitters. This may be a case in which you use microbore, because the pipes are as inconspicuous as electric cable. It may also, remember, be the right place to make those economical runs, of back-to-back radiators or other heat emitters, which we have described.

(6) By no means least and by no means last comes personal preference. Nobody has to give a reason for not liking gas, detesting oil or having no use for solid fuel — like the lady who refuses to have electricity in the place, except for lighting and the Hoover (which she claims are different).

The likes and dislikes can be more penetrating than that. A typical case in heating, though not in central heating, concerns the gas fire. With unbelievable perversity many interior design conscious people seem to object to the modern designed exterior of most gas fires and to look longingly at the gas fired log effect unit, which is one of the worst things to happen to the gas industry in this decade.

Manufacturers of boilers for gas, oil and solid fuel went to endless trouble to dress their kitchen boilers in a manner which makes them almost indistinguishable from the washing machine, the fridge, the kitchen unit. While most people approved, there remained a market for the bizarre design, to the delight of the few who retained it.

Most people have by now grasped the point that modern manufacturing methods, upon which we depend for a product of competitive price, cannot leave any room for the one-off. A unit has only one height, or width, a choice of only three colours, a flue at the back only − and so on. If you want something else it is useless to look for a variant of that brand. The correct thing to do is to look at another brand, even another manufacturer, in the hope of finding what you want. The quicker thing to do may often be to accept and adapt to the item which is not too welcome.

A particular case in which shopping around might pay is in radiators. There is often very good reason for wanting a radiator of a particular height, one which will fit a space available. But a manufacturer whose radiators are, say, 14 and 19 and 24 inches high has no possible chance of turning out one at 18 or 25 inches. So forget him, and look for the one in whose range 18 or 25 are standard. You may not care for the profile as much, but this becomes a matter of priorities.

Helping Hands

Almost everyone without any special knowledge of the subject seeks impartial advice. This is a natural reaction to the pressures exerted by advertising, and in particular by the advertising of fuels in contexts which are popularised from time to time. But we give it as a long considered opinion, that there is no such thing as impartial advice. Also that it is not such an ideal proposition as may at first appear. Let's be honest and admit that whenever there is a choice partiality creeps in. When we vote at an election, for instance. What we usually do is to try to find out what the issues are, before making up our own minds. And in that may be summed up the purpose of this book. When choosing a central heating system we expect that you will make your own mind up. All we aim to do is to put the main facts before you, to let you see the issues on which your decision should be made.

While we are trying to destroy long held beliefs, let us add that there is hardly ever a best system, for you or for anyone else. There is a short list of systems which have the greater number of the features which suit you and suit your premises, and in the long run it will matter hardly at all which of that list you choose. Indeed, the more thoroughly you follow the advice to insulate, the less important does any form of heating become.

We have tried to give you pointers to strong and weak points in a number of systems, knowing that these are not usually available in the publicity material. Overall, we have aimed to put you on at least nodding terms with what is available, so that you may turn to the handouts with more confidence, ask significant questions and evaluate the answers. But that does not mean that you are now to be cast adrift entirely on your own resources. Below is a list of some of the organisations which exist for your good. If you should need the services which they represent, in most cases the first thing to do is to find and approach the local representative. But if that is not satisfactory do not hesitate to go to the headquarters.

Certifying Authorities

For all gas fired appliances: British Gas. The local showroom might not stock what you have in mind, but they have a current list of approvals. Never use a gas appliance which is without British Gas approval.

All oil fired appliances: DOBETA. Domestic Oil Burning Equipment Testing Association Ltd., 3 Savoy Place, Victoria Embankment, WC2R OBN. Do not buy an oil fired appliance without DOBETA approval.

All solid fuel appliances: The Solid Fuel Advisory Service and Solid Smokeless Fuels Federation issue jointly a list of Approved Domestic Solid Fuel Appliances. Avoid anything not on that list. Your SFAS agent should have a copy, or the SFAS region office, or at Hobart House, London, SW1X 7AE.

All electrical appliances: BEAMA or Electricity Council approval, usually the Kite Mark (BSI) as well. If in doubt try the Electricity Council at 1 Charing Cross, London, SW1, but the local electricity showroom should be able to give the answers.

Radiators, convectors: MARC (Manufacturer's Association of Radiators and Convectors) certify heat ratings, and membership of the Association will be clearly stated on makers' literature, if it has been gained.

Water valves etc: are usually accepted by the National Water Council. But the basis of such approval is a British Standard, and it should therefore be quite satisfactory to establish that valves and fittings conform to the appropriate BS.

There are more specialist bodies, most of them of concern to the professional. Outside of that list most items are likely to be covered by customer protection legislation: Trades Description Act, and Sale of Goods (Implied Terms) Act, which give the customer much more power than formerly to demand an instant replacement for goods which do not do what is claimed.

We have mentioned British Standard's Kite Mark only briefly. This is no oversight. The organisations mentioned above are specialist and their function is particular to their calling. The Kite Mark must always be respected as an extra badge of merit, which takes in such features as continuing surveillance of manufacture and the maintaining of high standards. To contact BSI direct, the address is 2 Park Street, London, W1A 2 BS.

Doing the Work

One of the most potent fears is of getting into the clutches of a 'cowboy' and finding out too late that he is incompetent, or insolvent, or both. At the height of the boom, and the worst excesses, the late Heating Centre tried to run an insurance scheme against such an eventuality, but being short of money they were obliged to get the customer to pay the premium, which put the trade in the worst possible light. Since then more appropriate safeguards have grown up, and any customer who feels in need of protection should be able to get it. The most likely deterrent is still ignorance of the availability of such schemes.

For gas fired appliances there is CORGI, the Confederation of Registered Gas Installers. Most gas showrooms carry a list of CORGI approved installers, and if themselves engaged in installation work no doubt also belong. It should be noted that the only tested competence of CORGI members is in relation to the gas burning appliances. They are not tested for wet or dry systems at present, though one may go some way in assuming a willingness to seek approval. It is desirable therefore to make sure that you choose a CORGI man who happens to be experienced in the type of system you intend to have.

For all types of system there is a service offered by HVCA, the Heating and Ventilating Contractors' Association, with a double guarantee for customers. In this case the contractor pays the premium and HVCA underwrites the guarantee, and will arbitrate in case of dispute, and take over if the contractor goes bankrupt. The wide net cast by HVCA enables them to have servicing organisations on their lists too.

The way to activate HVCA is to tell them the type of system you are thinking of putting in, for example a wet system with oil fired boiler. They will then let you have a short list of members in the region who specialise in that kind of work, from which you will select two, perhaps three, and ask for a tender and specification. The latter is important if you are to choose between tenders, since you need to know exactly what each is offering for the money. The address to contact HVCA is Esca House, 34 Palace Court, Bayswater, London, W2 4JG.

You have a part to play in constructing the specification. It is your decision whether to have full heating, or partial, or any one of the choices listed in the Introduction. The choice of maintained temperatures too is yours. Then you can listen to any arguments in favour of, say, radiators versus skirting heaters; plain versus thermostatic radiator valves; small bore versus microbore; and so on. Remember that the installer or contractor is, in his own way, far from impartial. If he has reached your house because he is a specialist in one branch of domestic heating he is quite clearly partial to it. So do not be afraid to argue with him on any but strictly technical matters.

Getting a trustworthy and competent installer is the most important part of all decision making. If you do this you can practically forget about the need to conform to Building Regulations and all the rest of the list set out earlier. Your man will be fully aware of what he can and cannot do, who wants to know about what, and so on. He will almost certainly know which fuel tariff, in case of choice, is best suited to your new needs. For cavity wall insulation choose a manufacturer who belongs to the National Cavity Insulation Association, Bremar House, Sale Place, London W.2.

Costing

Although we have left this until last it is a very important factor in decision making, and the one which stays near the top of most people's lists. But in these troubled times we are not going to mention figures of any kind, but to draw comparisons and outline principles.

The question 'What will it cost?' is no more answerable than the other nonsensical 'How long is a piece of string?'. It depends upon almost everything. The proper objective for every householder is not to find out some fictitious datum cost but to make a determined effort to keep cost to a minimum compatible with the achievement of comfort. This is a two-part exercise. It entails first using the maximum of good insulation (Chapter 10): then running the system in accordance with the rules set out in the Introduction about quantity, place and time of warming.

That is what the cost of running a system comprises, and it will clearly vary widely (even after maximum insulation) according to number in family; whether home loving or always out, particularly out all day at work; whether young and active, or elderly and chilly.

The *real cost* of an installation, ignoring the effects of inflation for a moment, must be the first cost, what it cost to buy and install, plus the cost of running over a standard period of time, which is taken by some to be five years, by others as ten years. We can illustrate our point by looking at both.

Using wholly hypothetical figures, let us consider two installations. System A cost £500 to put in, and costs £150 a year to run, while system B cost £800 but runs for £100 a year. Which is the better proposition?

Over five years the cost of A is	£
Capital	500
Running: 5 x £150	750
Total for 5 years	1250
Cost per year	£250
The cost of B is	
Capital	800
Running: 5 x 100	500
Total	1300
Cost per year	260

Now taking a ten year period, similar calculations show that the annual cost of A is £200, that of B is £180.

We can see, then, that the annual cost falls as the influence of the first cost diminishes; also that for the same reason the cheaper running cost wins out in the end. In certain circumstances no doubt people may wish to amortise, i.e. spread the capital, over only five years. But if as is more usual the period chosen is a notional life of the system, then ten years is by no means unrealistic, and certainly favours a system which may cost more to buy but promises less cost to run. Incidentally that is not to suggest that a system will have only a ten year life. Most systems keep going long after the technology they represent has become a bit old fashioned. We cannot leave discussion of cost without a reference to real money, or the lack of it. With the cost of an installation now quite high, it is not everyone wanting a heating system who can or wants to pay cash.

For them there are hire purchase schemes, for instance run by the gas industry for a gas fired system, and perhaps allied to a particular type of system. It is always desirable to enquire, at the time of thinking of placing an order, what schemes are current available, for gas or oil or solid fuel or electricity and run by the fuel interests themselves — the Gas Region, the Electricity Board, the major oil company (Shell, Esso etc). Or it might be that one of the larger chains of builder's merchants operates a loan scheme. Major appliance manufacturers seem to become less implicated than formerly, and if offers are made from smaller or unknown suppliers one should take very great care to read the small print, and if necessary consult the bank manager, before signing anything.

Raising money by an extension of mortgage is still a possibility, and for shorter terms a bank loan is less expensive than it has been in recent times.

We have mentioned the greater protection afforded buyers nowadays by changes in the law. Let us conclude by linking this with the matter of guarantees. It is customary and proper for manufacturers of the sort of items with which we are concerned, boilers and radiators and pipes and so on, collectively called consumer durables, to give a period guarantee. In most cases it is a year. Sometimes, as recently with circulating pumps, a burst of selling activity persuades the period up to two years, even more. You cannot influence the period, but you can and should make sure that what you are buying has a stated time under guarantee, which dates from purchase and not from when it left the factory. And there must be a recognisable and acceptable way of showing as evidence the date of purchase, which can be registered with the maker or the supplier.

The one thing the law has done is to render meaningless those so-called let-out clauses by which a manufacturer may seek to impose his own conditions and deprive the customer of any rights under common law.

If asked to sign anything you may safely strike them out, though if left in they have no weight.

The one way in which the manufacturer is able, rightly, to deny any responsibility under his guarantee is in the case of tampering with the product so as to upset its normal functioning or construction. The manufacturer, in short, expects and has a legal right to expect, that when he makes a technical product it shall be handled at all stages by persons competent to do so. When we carry this to the ultimate point, the user, it would of course be quite unrealistic to expect a general expertise. In consequence all the complications are hidden away, and the user is left with a knob, a switch, a push button, or a small collection of such items, and clear instructions. If something goes wrong with those elementary items the manufacturer will admit responsibility. But if the customer attempts to do his own repairs, and upsets the more complex arrangements which are hidden away, then he is likely to forfeit any rights under a guarantee. Nobody should quarrel with this, and it is indeed something of a cause for wonder, that as the technology becomes more complex, and more exacting, it can be so successfully hidden, yet still brought to the surface in the most elementary devices for the use of the user.

It does no harm to contemplate occasionally, just how much the user and the manufacturer have in common. For purely selfish motives of profit the manufacturer wants the customer to get the maximum of satisfaction and the minimum of trouble from the appliance, whatever it is. And isn't that exactly what the customer wants? It is our turn to wish you the same — maximum satisfaction and minimum trouble. Our motive is a belief in good central heating which we hope will be catching.

Index